Introduction to Python

Introduction to Python: with Applications in Optimization, Image and Video Processing, and Machine Learning is intended primarily for advanced undergraduate and graduate students in quantitative sciences such as mathematics, computer science, and engineering. In addition to this, the book is written in such a way that it can also serve as a self-contained handbook for professionals working in quantitative fields including finance, IT, and many other industries where programming is a useful or essential tool.

The book is written to be accessible and useful to those with no prior experience of Python, but those who are somewhat more adept will also benefit from the more advanced material that comes later in the book.

Features
- Covers introductory and advanced material. Advanced material includes lists, dictionaries, tuples, arrays, plotting using Matplotlib, object-oriented programming
- Suitable as a textbook for advanced undergraduates or postgraduates, or as a reference for researchers and professionals
- Solutions manual, code, and additional examples are available for download.

Chapman & Hall/CRC

The Python Series

About the Series

Python has been ranked as the most popular programming language, and it is widely used in education and industry. This book series will offer a wide range of books on Python for students and professionals. Titles in the series will help users learn the language at an introductory and advanced level, and explore its many applications in data science, AI, and machine learning. Series titles can also be supplemented with Jupyter notebooks.

Image Processing and Acquisition using Python, Second Edition
Ravishankar Chityala and Sridevi Pudipeddi

Python Packages
Tomas Beuzen and Tiffany-Anne Timbers

Statistics and Data Visualisation with Python
Jesús Rogel-Salazar

Introduction to Python for Humanists
William J.B. Mattingly

Python for Scientific Computation and Artificial Intelligence
Stephen Lynch

Learning Professional Python Volume 1: The Basics
Usharani Bhimavarapu and Jude D. Hemanth

Learning Professional Python Volume 2: Advanced
Usharani Bhimavarapu and Jude D. Hemanth

Learning Advanced Python from Open Source Projects
Rongpeng Li

Foundations of Data Science with Python
John Mark Shea

Data Mining with Python: Theory, Applications, and Case Studies
Di Wu

A Simple Introduction to Python
Stephen Lynch

Introduction to Python: with Applications in Optimization, Image and Video Processing, and Machine Learning
David Baez-Lopez and David Alfredo Báez Villegas

For more information about this series please visit: https://www.crcpress.com/Chapman-
-HallCRC/book-series/PYTH

Introduction to Python

With Applications in Optimization, Image and Video Processing, and Machine Learning

David Báez-López
David Alfredo Báez Villegas

CRC Press
Taylor & Francis Group
Boca Raton London New York

CRC Press is an imprint of the
Taylor & Francis Group, an **informa** business

A CHAPMAN & HALL BOOK

Designed cover image: ShutterStock Images

First edition published 2024
by CRC Press
2385 NW Executive Center Drive, Suite 320, Boca Raton FL 33431

and by CRC Press
4 Park Square, Milton Park, Abingdon, Oxon, OX14 4RN

CRC Press is an imprint of Taylor & Francis Group, LLC

ISBN: 978-1-032-11767-6 (hbk)
ISBN: 978-1-032-11910-6 (pbk)
ISBN: 978-1-003-22211-8 (ebk)

DOI: 10.1201/9781003222118

Typeset in CMR10 font
by KnowledgeWorks Global Ltd.

Publisher's note: This book has been prepared from camera-ready copy provided by the authors.

Dedicated to Laura and Gary, my children, and to Alina, my granddaughter.
The future is yours.

For David, Lucero, Fidel, and Eric.
Who have made everything I have achieved possible.

Contents

Preface

This book was born from a programming course taught at Universidad de las Americas-Puebla, Mexico. In that time, Python was gaining strength as a powerful programming language, but nobody foresaw that it could become one of the most used programming languages. In fact, the TIOBE index ranks Python as the most used programming language in the world (April 2024). This fact is due to the characteristics it possesses. Python is a high-level programming language that supports procedural, imperative, object-oriented, and functional modes. This book covers the first three modes. Python has a simple easy to learn syntax. Python supports modules, libraries, and packages, encouraging program modularity and code reuse. Python and the majority of supporting libraries are open-source, a characteristic that is appealing for designers and users. This fact has enriched the modules and tools freely available. In fact, Python and most of the libraries available are open-source. There are libraries for science, engineering, finance, mathematics, statistics, machine learning, data analysis, and many more.

The book may be divided in two parts. The first part, comprised of Chapters 1 to 8, can be used by a person interested in learning Python from scratch; that is, he/she does not have any knowledge of Python. The learning goes from installing Python to implementing functions and generating complex plots. The second part covers advanced topics such as optimization, image and video processing, and machine learning using Keras, TensorFlow, and neural networks. Throughout the book, the use of external modules or libraries makes the Python programs very powerful.

The book starts with basic Python instructions and introduces the concept of algorithmic solutions. The basic concepts in a Python program are presented. The second chapter is about conditionals and loops. The availability of these tools are of paramount importance in any programming language.The concepts of nested conditionals and loops are presented. These instructions allow the implementation of complex algorithms. The concepts of data structures in Python is covered in Chapter 3. The data structures are covered in detail together with their methods. Arrays are covered next. They are defined and used in matrix algebra. Arrays from Pandas are presented. Chapter 5 treats functions and their use in larger programs. Chapter 6 presents the object-oriented programming paradigm. The different properties that are associated are described. Reading data from files and writing data to files is presented in Chapter 7. The visualization of data is an important part of modern programming languages and Python has associated a library that permits to create complex plots, either in two dimensions or in three dimensions. Another package called Seaborn is also used in the generation of graphs. The first eight chapters can be used by a novice student to gain a good expertise in Python programming.

The second part of the book is four chapters long and it covers concepts of optimization using Scipy that has a library of functions to perform optimization. Several examples show the techniques used. Image and video processing are implemented using OpenCV, a package especially designed to handle images and video. Machine learning is the topic of the last two chapters. Chapter 11 implements supervised and unsupervised learning using Keras and Tensor-Flow. Chapter 12 makes use of neural networks to implement machine learning algorithms. The version of Python used in the book examples was Python version 3.12.

Finally, the authors wish thank the staff at Taylor and Francis for the support to take this project to a happy ending. Special thanks Callum Fraser and Mansi Kabra for the encouragement to complete the project.

David Báez-López
David Alfredo Báez Villegas

Cholula, Mexico, March 2024.

Author Biographies

David Báez-López obtained a B.S. in Physics in 1973 from the Universidad Autonoma de Puebla, Mexico where he graduated with Honors. He also attended the University of Arizona where he obtained the M.S. and Ph.D. degrees, both in Electrical Engineering, in 1976 and 1979, respectively. From 1979 to 1984 he was a researcher at Department of Electronics at National Institute for Astrophysics, Optics, and Electronics, located in Tonantzintla, Puebla, Mexico, and he became Head of The Department of Electronics and Graduate Coordinator. In 1985 he joined Universidad de las Americas-Puebla, Mexico as a professor and in 1988 became Chairman of the Department of Electronic Engineering until 1994. He was a member of the National Systems of Researchers. He retired in 2015 and now is Academic Director in EDUPROTEC A.C. in Puebla, Mexico, a non-governmental organization dedicated to the improvement of mathematics learning, among other topics. He has authored books on circuit simulation, MATLAB, Python, and Mathematica in English and Spanish. A Chinese edition of a book on MATLAB was published in 2017.

David Alfredo Báez Villegas is an electrical engineer with master's degrees in Project Management (École de Technologie Supérieure, Montréal) and Electrical Engineering (Texas A&M University, Texas Tech University). He holds a B. S. in Electronic Engineering from Universidad de las Americas-Puebla, Mexico, and M.S in Electrical Engineering from Texas Tech University and from Texas A&M University (TAMU). He worked as an assistant to the Office of International Affairs at TAMU before moving to Montreal, Canada. Since 2015 he has been a consultant in project management.

Chapter 1

Introduction to Python

1.1 What is the Python Programming Language

This book is about the Python programming language. This is a high-level programming language. In April of 2024, Python ranked as the most used language among many others such as C, C++, MATLAB, Java, and JavaScript, just to mention a few.

When a program is written in any programming language, what the user writes is called the source code. This code has to be converted to machine language code, and for this purpose, a compiler is used. The result is a file known as the object code. Finally, from this object code the compiler produces the executable file and this file is independent of the language used.

Other programming languages compile the source code on an instruction-by-instruction basis. These types of compilers are known as interpreters.

A task that the compiler does is to check the syntax of the source code. If there are any errors, the compilation process indicates all the errors in the source code. If there are no errors, the executable file is created. In the case of the interpreters, an instruction is checked, and if there is an error, the process is interrupted to correct the error. The disadvantage of an interpreter is that if there are 10 errors in the program, then 10 compilations and 10 corrections have to be made.

Python uses an interpreter to check the syntax. This fact makes the compilation process slower as compared to other programming languages such as C, C++, and Java, to name a few. However, it is possible to create an executable file that runs faster as compared to the interpreter.

1.2 The Python Programming Language

Python was introduced in its version 0.9.0 in 1991. It was conceived by Guido von Rossum. It is a high-level language and is open source. The most used versions are Python 2 and Python 3. In this book Python 3 is used.

DOI: 10.1201/9781003222118-1

1

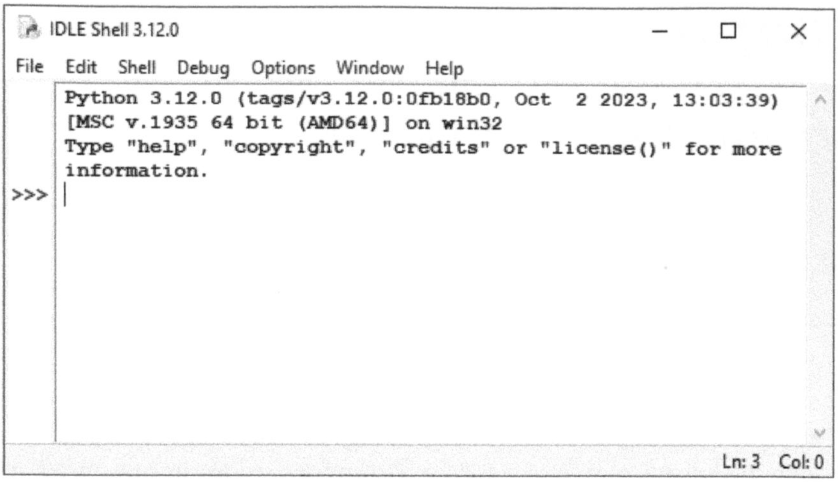

Python with Optimization, ...

```
IDLE Shell 3.12.0                                   —    □    ×
File   Edit  Shell  Debug  Options  Window  Help
       Python 3.12.0 (tags/v3.12.0:0fb18b0, Oct  2 2023, 13:03:39)  ^
       [MSC v.1935 64 bit (AMD64)] on win32
       Type "help", "copyright", "credits" or "license()" for more
       information.
 >>>   |

                                                        Ln: 3  Col: 0
```

FIGURE 1.1: Python's IDLE.

Python is platform independent and that is an important advantage. A Python program written in Windows can be used in Linux or Mac computers, and vice versa. The only requirement for this is that the program is written in the same Python version.

Python has a standard library of functions. This fact allows programmers and designers a great deal of flexibility in producing very powerful code. In addition, users have developed modules and libraries, most of them open-source and free. Those modules can be used to develop graphical interfaces, plotting, develop financial, mathematical, and engineering applications, to name a few.

1.2.1 Downloading Python

Python can be downloaded from The Python Software Foundation page at **www.python.org**. The version used in this book is the most recent version, version 3.12 at the beginning of 2024. To install it, the instructions indicated in Appendix A need to be followed. It is important to install on disk C: for windows platforms. Once installed, it is ready to use.

1.2.2 The Python's Integrated Development and Learning Environment

Python has an Integrated Development and Learning Environment (IDLE), where variables and functions are defined and executed. Also, results for scripts and programs are displayed at the IDLE. Figure 1.1 shows the IDLE. In this figure, the prompt is shown with the symbols >>>. It can be seen that the Python version has a word length of 64 bits.

1.3 Book Organization

The book is designed to be read by students and programmers with no previous experience in Python programming. It can also be used by people with previous knowledge of Python and who require a deeper knowledge of it. In any case, the examples in this book as well as the exercises at the end of the chapters help to improve the skills in Python programming. The book can also be used by expert Python programmers to learn the topics on image and video processing, optimization, and machine learning, topics that have a great deal of applications.

Chapter 1 is introductory to Python. Operations are made from the IDLE, also known as the Python shell. From there, readers learn to make simple programs, known as scripts. The user learns to use basic functions by importing modules and libraries of functions.

Chapter 2 treats the concepts of conditionals and loops which serve to control the flow of a program.

Chapters 3 and 4 give a treatment for data structures such as strings, lists, dictionaries, tuples, vectors, and matrices.

Functions are the subject of Chapter 5. In this chapter the ways functions can be defined is covered.

Data read and written to files is covered in Chapter 6.

Chapter 7 presents the Object-Oriented Programming (OOP) paradigm.

Data visualization is covered in Chapter 8 where `matplotlib` is used to obtain high-quality plots.

Python can be used for image and video processing. This topic is covered in Chapter 9.

Optimization using Python and Scipy can be used to optimize functions. This is the topic of Chapter 10.

The last two Chapters, 11 and 12, are dedicated to the use of Python in machine learning, which include neural networks using Scikit-Learn and TensorFlow.

1.4 Algorithms

An algorithm is a way to solve a problem. Examples of algorithms in everyday life can be found in cooking recipes, instructions to drive to a place, how to assemble pre-built furniture, fixing a computer by a technician, etc. In all of these examples, there is a sequence of steps to successfully complete the required assignment. The last step in the design of the algorithm is the testing step to insure that the algorithm behaves as expected. A more accurate definition of an algorithm is:

An algorithm is a precise finite sequence of steps to solve a problem and to achieve a result.

The necessary steps to implement an algorithm are:

1. Analysis of the problem.
2. Design of the algorithm to solve it.
3. Verification of the algorithm.
4. Implementation of the algorithm in some programming language.
5. Testing of the algorithm with known data.

The word algorithm is to honor the Arab mathematician Al Khuarismi who is believed to be the creator of the first algorithm to obtain the roots of an equation.

1.5   Variables

Variables are data that can take on different values depending upon the algorithm. Variables must have a name to identify them. The name is a string of alphanumeric characters which must start with a letter. The name may have numbers and underscores, either capital or lower-case letters. Also, the name is unique; that is, no other variable can have the same name. In Python, a capital letter is different from a lower-case one; thus, the variable A1 is different from a1.

In every programming language, there are reserved words called *keywords* and cannot be used as variable names. Python 3 has 33 keywords. To display them in the IDLE type:

```
>>> import keyword
>>> keyword.kwlist
```

and the result is:

['False', 'None', 'True', '__peg_parser__', 'and', 'as', 'assert', 'async', 'await', 'break', 'class', 'continue', 'def', 'del', 'elif', 'else', 'except', 'finally', 'for', 'from', 'global', 'if', 'import', 'in', 'is', 'lambda', 'nonlocal', 'not', 'or', 'pass', 'raise', 'return', 'try', 'while', 'with', 'yield']

Names of variables must start with a letter and can only have letters, numbers, and underscores.

1.5.1 Types of Variables

In Python, and in any other programming language, each variable has a certain type. The most used variable types in Python are floating-point, integer, alphanumeric or text, and boolean.

A variable is a **floating-point** variable if it has a decimal point and an integer part to the left of the decimal point, and a fractional part to the right of the decimal point. The fractional part is known as the mantissa. Thus, floating-point variables are of the form:

12345.987

A variable is an **integer** one if it does not have a fractional part; that is, it only has an integer part. For example, $a = 6$, `speed = 45`, `depth = -3` are integer variables.

An **alphanumeric** variable or text variable is a string of alphanumeric characters which may include letters, numbers, and any other character. These variables are called **strings** and are enclosed in single or double quotes such as "`elephant`", '`silver_coin`', '`year 2025`'. The last string includes a blank space which is counted as a character.

A **boolean** variable is a variable that can only take the values `True` or `False`. They are used whenever a decision is taken. Boolean variables are named after George Boole, a British mathematician who invented what is now called Boole's algebra.

The type of the variable is assigned when the variable is defined. Thus, the expression $a = 3$ defines an integer variable, $b = -3.24$ defines a floating-point variable, $c =$ `True` defines a boolean variable, and $d =$ "`animal`" defines a string.

Python is a *dynamic-type* language. This means that a variable of a certain type can be redefined in a program with another type. For example, the variable $d =$ "`animal`" can be redefined in the program as $d = -0.32$, a floating-point variable, and later on redefined as $d =$ `True`, a boolean variable.

To find out what is the type of a variable the `type` instruction is used. For example, if variables a, b, c, and d are given by:

```
>>> a = 2
>>> b = 5.6
>>> c = True
>>> d = 'ENIAC'
```

The type of these variables is then:

```
>>> type(a)
<class 'int'>
>>> type(b)
<class 'float'>
>>> type(c)
<class 'bool'>
>>> type(d)
<class 'str'>
```

It can be seen that a is an integer variable, b is a floating-point one, c is a boolean variable, and finally, d is a string.

The type of any variable can be changed by using the functions int to change to integer, float to change to floating-point, bool to change to boolean, and str to change to string.

Example 1.1 Change of type for a variable

The variable a = 9 has an integer type. The type of this variable can be changed as:

```
>>> float(a)
1.0
>>> bool(a)
True
>>> str(a)
'1'
```

1.5.2   Variable Assignment

The value for variables a1 and a2 are assigned as:

```
a1 = 4
a2 = 8
```

In this way, a1 has the value 4 and a2 has the value 8. It is said that the value 4 is assigned to a1 and the value 8 is assigned to a2. This process is called an **assignment**. If now the following is done:

```
a1 = a2
```

Now a1 is assigned the value of a2; thus, the new value of a1 is 8. The old value of 4 is lost. The value of a2 has been assigned to a1.

1.5.3 Basic Operations

Basic operations are implemented in a traditional way in the IDLE. Usually, the result is assigned to a variable. For example:

```
>>> a = 3
>>> b = 5
>>> a + b
```

gives a result:

```
8
```

If a new variable is defined, the result is not automatically displayed. The user must write the variable name at the Python prompt, as:

```
>>> c = a + b
>>> c
```

and the result is displayed:

```
8
```

1.6 Input and Output in Python

In the Python IDLE and in programs, the input of data is done using the instruction **input**. This instruction reads data as strings of data. Thus, it is natural to read in text. However, most of the data to be processed in Python programs or scripts is numerical data and the strings read in must be converted. This conversion can be done by changing the type of the strings. For example, to read a floating-point number use:

```
x = input( )
x = float(x)
```

These two instructions can be combined in a single row by assigning the type to the string read with the input instructions. For example:

```
x = float(input( ))
```

The argument of the **input** instruction can have a text between double or single quotation marks to indicate users what data is being read, as in:

TABLE 1.1:   Format options

Symbol	Output type	Example
d , i	Signed integer	-123
u	Unsigned integer	123
f	Floating-point	123.45
e	Scientific notation	1.2345 e+002
g	Shortest representation between f and e	123.45
c	Character	'a'
s	String	"program"

```
>>> g = float(input("Enter the value of gravity:  "))

Enter the value of gravity: 9.80
```

This indicates that the variable **g** is a floating-point one and its value is g = 9.8.

For the output of data, the instruction **print** can be used. To display the value of **g** above, it can be done with:

```
print(g)
```

This instruction can also have a text in it as:

```
print("The value of gravity is: ", g)
```

A format can be added inside the instruction **print**. The format options are shown in Table 1.1. A formatting instruction can be added inside a **print** instruction as in this example for the gravity as:

```
>>> print("The value of gravity is %0.3f:  ", g))
```

that displays:

```
The value of gravity is 9.800
```

The use of a **print** instruction provides a line feed. If there are two consecutive **print** instructions, the second **print** instruction is displayed in the next line. For example,

```
>>> print("The value of gravity is :  ")
>>> print(g)
```

produces:

```
The value of gravity is :
   9.8
```

If it is desired to display in the same line, an end-line sign must be specified. This end-line sign is **end** = " ". These instructions must be within a script. For the variables **a** = 4, and **b** = -7 as:

```
print("The value of a is: ", a, end = ". ")
print("The value of b is: ", b)
```

The data is displayed as:

```
The value of a is: 4. The value of b is: -7
```

Formatted string literals, called **f-strings**, can include the value of a Python expression by using the prefix **f** and writing the expression between braces. The last example can be written as

```
a = 4
b = -7
print(f'The value of a is: {a}. The value of b is: {b}')
```

which produces the same result.

Python uses the Unicode standard which is a standard representation for characters and strings. This lets Python programs work with all these different possible characters. The Unicode standard is available at https://www.unicode.org. For example, the number π, which is available in the library **math**, has the Unicode \u03C0. To print out the value of the number π the following is used:

```
from math import pi
>>> print("The value of \u03C0 is %0.10f: ", pi))
```

Libraries or modules are explained later in the chapter.

1.6.1 Escape Sequences

An escape sequence is a sequence of characters used inside a string. All escape sequences consist of two or more characters and in every case the first character is a backslash \. The following characters determine the interpretation of the escape sequence. Some of the most used escape sequences are shown in Table 1.2.

TABLE 1.2:   Escape sequences

Escape sequence	Meaning
\b	Backspace
\f	Formfeed Page Break
\n	Newline (Line Feed)
\r	Carriage Return
\t	Horizontal Tab
\v	Vertical Tab
\'	Apostrophe
\\	Backslash

Example 1.2 Escape sequences

An example of escape sequences is in the use of the **print** instruction. Suppose the variables a = 1, b = 2, and c = 3 are to be printed and separated by the same distance using a horizontal tab escape sequence. Then, the **print** instruction is:

```
>>> print(a, '\t', b, '\t', c)
```

and the data displayed is:

```
1     2     3
```

However, if the data is to be written in different rows, then a newline escape sequence is used and the **print** instruction is:

```
>>> print(a, '\n', b, '\n', c)
```

and the result is:

```
1
2
3
```

1.7   Programs in Python

To perform complex computations a program is written in Python. This program is stored in a separate file and run from the Python IDLE. Programs in Python are also referred to as scripts. An example of a program is the script to solve a first-degree equation of the form:

$$ax + b = c$$

that can be solved as

$$x = \frac{c - b}{a}$$

This solution can be programmed[1] as:

```
a = float(input("Enter the value of coefficient a: "))
b = float(input("Enter the value of coefficient b: "))
c = float(input("Enter the value of coefficient c: "))

x = (c - b)/a
print("The solution of the first-degree equation is x = ", x)
```

To create this script, in the menu bar in the Python IDLE select

```
File → New File
```

as shown in Figure 1.2. This opens a new window where the program can be written. Before executing the script, it must be saved to a convenient location with the name: `Solution_first_degree_equation.py`. To run this program, in the menu bar of the program window, select `Run → Run Module` as shown in Figure 1.3. The result is:

```
Enter the value of coefficient a: 2
Enter the value of coefficient b: 12
Enter the value of coefficient c: 16
The solution of the first-degree equation is x = 2.0
```

1.8 Comments in a Program

Comments in a program are of paramount importance because they help programmers and users to understand the code. They can be implemented at the beginning and in between lines of code. Comments allow users to understand the program and they can be used to more easily debug the program. Important characteristics of comments are: 1. They must be short giving precise information about the purpose of the program. 2. They must add informative value. There are single-line comments that start with a hash character # followed by a comment. Every line of comments must start with a hash character. There are comments that explain a line of code and they may be written after

---

[1]Programs or scripts are enclosed in a box.

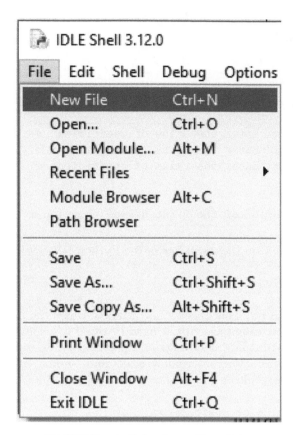

FIGURE 1.2: Selecting a New File.

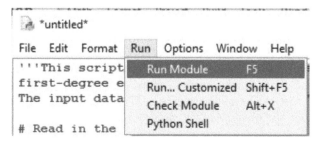

FIGURE 1.3: Selecting a New File.

the code starting with a hash character. Finally, there are multiple lines of code. These lines of code start with three single or double quotation marks to open the lines of code. Triple single or double quotation marks must close the comment block. For the solution of the first-degree equation, comments may be implemented as:

```
"'This program finds the solution of a first-degree equation.
The equation is of the form ax + b = c.
The solution is of the form: x = (c - b)/a '"

# The coefficients are read:
a = float(input("Enter the value of coefficient a: "))
b = float(input("Enter the value of coefficient b: "))
c = float(input("Enter the value of coefficient c: "))

# The solution is computed:
x = (c - b)/a

# The result is displayed:
print("The solution of the first-degree equation is x = ", x)
```

1.8.1 Operations with Integer and Floating-point Numbers

When one of the variables is an integer one and the other variable is a floating-point number the result is a floating point number. For example, for a = 3.4 and b = 2, then the results are:

```
>>> a = 3.4
>>> b = 2
>>> x = a + b
>>> print(x)
5.4
```

1.9 Functions in Python

The concept of function is associated with a method to do something. The first contact of a student with functions is with the trigonometric functions in high school, but in real life there are many examples.

TABLE 1.3:   Built-in functions

Function	Python	Meaning
a^b	pow(a, b)	Power
quotient of a/b	a//b	Obtains the integer part of a/b
residue of a/b	a%b	Obtains the residue of a/b
a/b	divmod(a,b)	Obtains quotient and residue of a/b
abs()	abs(a)	Absolute value of a
max()	max(a, b)	Maximum of a and b
bin()	bin(a)	Binary representation of a
hex()	hex(a)	Hexadecimal representation of a

In the Python IDLE, computations can be made and functions can be run. As it was indicated before, the Python IDLE is ready to receive data when the prompt appears. A list of some built-in basic functions available in Python is shown in Table 1.3. This list is not exhaustive and a complete list is available at https://docs.python.org/3/library/functions.html. Chapter 5 covers the instructions that let users create functions.

1.10   Modules and Libraries

In the basic Python shell, there are available basic functions. Although the number of built-in functions is limited in Python, there are other modules or libraries available to use them. Some of the modules available that are used in the book are math, random, matplotlib, NumPy, SciPy, Scikit-Learn, and TensorFlow. The modules have to be imported before being used. Each module or library of functions has subsets that have to be imported as part of the larger library. The modules or libraries will be referred to indistinctly as modules or libraries. While some are available when Python is installed, like the modules math and random, some other modules or libraries require an installation as described in Appendix A. Once installed, the libraries can be imported and used.

There are three ways to import a function. The three methods are described here and to show them the square root function is used.

1. The square root function is available in the library math. To import a library and use the square root function use:

```
import math
a = 3
b = math.sqrt(a)
```

In this case each and everyone of the functions available in the module are imported. This makes the program larger.

2. Instead of importing the whole module, import only the square root function. This is done with:

```
from math import sqrt
a = 3
b = sqrt(a)
```

This case is optimum because only the function required is imported. It is not necessary to indicate that the function sqrt belongs to the library math.

3. If several functions of a module are going to be used, then use:

```
from math import *
```

In this case, every function in the module is imported.

A common way to import modules or libraries is associating a shorter name, for example, to import numpy the following can be used:

```
import numpy as np
```

and from there on use np, for example:

```
b = np.sqrt(a)
```

To learn which functions are available in a module or library, import the module and then use dir(module_name). For example, for the library numpy:

```
>>> import numpy
>>> dir(numpy)
```

The option to display 92 lines is printed in the IDLE. Clicking on the option displays all the functions available in the library numpy.

Python with Optimization, ...

TABLE 1.4:   Arithmetic operators

Operation	Operator	Example	Precedence
Addition	+	a + b	3
Subtraction	-	a - b	3
Multiplication	*	a*b	2
Division	/	a/b	2
Integer division	//	a//b	2
Power a^x	pow	pow(a, x)	1

1.11   Operators

Operators indicate the type of operations that can be performed on the variables. There exist four types of operators:

1. Arithmetic.

2. Relational.

3. Logic.

4. Assignment.

1.11.1   Arithmetic Operators

Arithmetic operators require two arguments, although sometimes only one argument is required. Table 1.4 gives a list of the arithmetic operators.

The result of an operation is of the same type as the type of the arguments if both are of the same type. In case an operator is a floating-point one and the other one is an integer, the result is a floating-point one.

1.11.2   Relational Operators

Relational operators, also called comparison operators, are used when a comparison must be done. They are extensively used in the next chapter. Table 1.5 gives a list of the relational operators. Relational operators are used to compare and the comparison produces a boolean result.

TABLE 1.5:   Relational operators

Operation	Operator	Example
Greater than	>	a > b
Less than	<	a < b
Greater than or equal to	>=	a>=b
Less than or equal to	<=	a<=b
Equal to	==	a == x
Different from	!=	a != x

TABLE 1.6: Logical operators

Operation	Operator	Example	Precedence
And	and , &	a and b, a & b	2
Or	or, |	a or b, a | b	4
Not	not	not a	1
Xor	∧	a ∧ b	3

Example 1.3 Relational operators

Given the variables a = 3 and b = 6, the following use of relational operators produces the indicated results:

```
a < b  → True
a > b  → False
a == b → False
a != b → True
```

It can readily be seen that the these statements can only have a Boolean result, namely, either `True` or `False`.

1.11.3 Logical Operators

The logical operators are used with boolean variables, also known as logical variables. Boolean variables can only have the values `True` and `False`. Table 1.6 shows the main logical operators and their precedence.

Example 1.4 Logical Operators

Given the variables a, b, and c, the following operations can be done:

```
>>> a = True; b = True; c = False
>>> a & b
True

>>> a and b
True

>>> a and c
False

>>> a or c
True

>>> a|c
True
```

TABLE 1.7:   Assignment operators

Python	Operator
a = b	=
a = a + 1	a += 1
a = a − 1	a −= 1
a = a + b	a += b
a = a − b	a −= b
a = a * b	a *= b
a = a / b	a /= b
a = a % b	a %= b

```
>>> a ∧ b
False

>>> a ∧ c
True
```

It is convenient to use parenthesis and then first evaluate the parenthesis and then the remaining operations using precedence. For example, if a = False, b = True, and c = True, then

$$a \ \& \ b \ | \ c$$

has the result True because a&b = False and False|c is True, but

$$a\&(b|c)$$

is False because first (b|c) is computed and the result is True, but a&True is False. Thus, using parenthesis alters the order and evaluation and the result.

1.11.4   Assignment Operators

Assignment operators assign or modify the value of a variable. Some very useful assignment operators are shown in Table 1.7.

Example 1.5 Assignment operators
For the variables a = 5, b = 3, the following results are computed:

```
a += 1 → a = 6
a −= 1 → a = 4
a += b → a = 8
```

```
a -= b → a = 2
a *= b → a = 15
a /= b → a = 1.66667
a %= b → a = 2
```

1.12 Alphanumeric Variables

Alphanumeric variables in Python are known as strings. Strings are enclosed in single or double quotation marks, for example, "dog", 'cat', '2702', 'February 28', 'January_14', 'floating-point'. As can be seen, a string can contain underscores, blank spaces, and any other character available. The type of a string is `str`. Strings have already been used in the `print` and `input` instructions together with escape sequences. Strings are covered in Chapter 3 and here only a few functions are covered.

The length of a string can be obtained with the function `len()`. This gives the number of characters in the string including blank spaces. For example:

```
>>> len('February 28')
11
```

The result is 11 because the blank space is also counted. Strings can be concatenated, or added, with the symbol $+$. For example, given the strings a = 'cats', b = 'dogs', c = ' ', d = 'and', they can be added as:

```
>>> a + c + d + c + b
cats and dogs

>>> len(a + c + d + c + b)
13
```

The blank string is used for separation of the strings. The length of the string is 13 because the two blank spaces are also counted.

Each character in a string has a position. The numbering of the position may start from the left and in that case, the left-most character is in the position 0, the next element to the right is in position 1, and so on. The last element in a string with length n is in position n-1. If the position numbering starts from the right, then the right-most element is in position -1, the next element to the left is in position -2, and so on. The left-most element is in position -n. The numbering of a string is depicted in Figure 1.4.

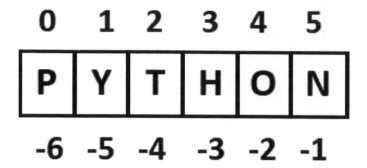

FIGURE 1.4: Position of characters in a string.

1.13 Lists

Sometimes it is necessary to work with a set of data. In Python, a list is such a set. A list of elements is enclosed in brackets and the elements of the list can be of several types, for example, some elements can be integer ones, but other elements can be floating-point elements, strings, or even lists. Elements in a list must be separated by a comma. A list must have a finite number of elements. As an example, a list of the first ten natural numbers is:

```
natural = [1, 2, 3, 4, 5, 6, 7, 8, 9, 10]
```

Another example is a list of friends with their age; thus, the list elements can be strings and integer numbers:

```
>>> friends = ['Pete', 16, 'Rose', 15, 'John', 21, "Louis", 19]
```

The number of elements of a list can be found with the function `len()` as was the case for strings. Further, lists can be concatenated and multiplied by an integer. For example,

```
>>> name = ["Tony", 17]
>>> friends + name
['Pete', 16, 'Rose', 15, 'John', 21, 'Louis', 19, 'Tony', 17]

>>> name*3
['Tony', 17, 'Tony', 17, 'Tony', 17]
```

The instruction **len** is used to find out the number of the list **name** as:

```
>>> len(name)
2
```

because there are only two elements in the list **name**. But if it is desired to find the number of elements in the first element of the list **name**, the result is:

```
>>> len(name[0])
4
```

because there are four elements in the string "Tony".

The position of the elements in a list follows the same rules of strings. In the list **dogs** = ["Cocker", "Grayhound", "Dalmatian"], "Cocker" is in the position 0, , "Grayhound" is in the position 1, and "Dalmatian" is in the position 2. Alternately, , "Dalmatian" is in the position -1, "Grayhound" is in the position -2, and "Cocker" is in the position -3.

1.14 Dictionaries

Dictionaries are lists whose elements are pairs of variables. Each pair is composed of a key and a value. The key and its value are separated by a colon. The pairs are separated by commas. For example,

```
>>> dictionary1 = { 'a' : 23, 'moon' : 'unique', 'x' : 7.3 }
```

To print the value of a pair only the key is needed. To print the value of key **a** in **dictionary1** use:

```
>>> print("dictionary1['a']: ", dictionary1['a'])
dictionary1['a']:23
```

which is the correct result.

The length of a dictionary is obtained with the instruction **len**. For example, consider the following dictionary whose length is going to be found:

```
>>> mat = { "Algebra":[20, 30, 40] , "Arithmetic":[17, 41] }
>>> len(mat)
2
```

The answer is 2 because in the dictionary `mat` there are 2 pairs. Note that in the dictionary `mat` the values are lists.

The list `friends` from the previous section can be better written as a dictionary. This can be done with:

```
friends = {'Pete': 16, 'Rose': 15, 'John': 21, 'Louis': 19, 'Tony': 17}
```

Dictionaries are not ordered and, therefore, there is no position associated with the pairs. Thus, as it was stated above, only the key is needed to display a value in an element of a dictionary.

1.15 Tuples

A tuple is an ordered collection of elements. Tuples may or may not be enclosed in parentheses. The elements of a tuple can be strings, integers, floating-point, lists, dictionaries, or any combination of elements. The length of a tuple can be found with the function `len( )`. For example, the tuple `animals = ("cats", "dogs", "birds")`, the length can be found with:

```
>>> animals = ("cats", "dogs", "birds")
>>> len(animals)
3
```

Concatenation and multiplication by an integer can also be done, as was the case for strings, lists, and dictionaries.

The position of the elements in a tuple follows the same rules of strings.

1.16 Examples

This section presents three examples of small programs or scripts in Python. The first example uses tuples to swap variables, the second example computes simple and compound interest and the last example implements temperature conversion. The scripts have to be written in a new file and run as described in Section 1.7.

Example 1.6 Variable swap
Sometimes variables need to be swapped. Usually, the swap requires a temporary variable. For example, to swap the variables x and y, the following is done:

First, save the value of x in a temporary variable:

```
>>> temp = x
```

Second, store the variable y into variable x:

```
>>> x = y
```

Finally, store the temporary variable which has the value of x into y:

```
>>> y = temp
```

The final result is that the values of the variables x and y have been swapped. There is another way to do the swapping using tuples and the whole swapping is done with:

```
>>> (x, y) = (y, x)
```

For example, given the tuple (-7, 23), the values are swapped with:

```
>>> (x, y) = (-7, 23)
>>> (x, y) = (y, x)
>>> print("The value of x is %d: '%x)
```

The value of x is 23:

```
>>> print("The value of y is %d: ' %y)
```

The value of y is -7:

Thus, the swapping has been done.

Example 1.7 Temperature conversion
Conversion of temperatures in the three scales Celsius, Fahrenheit, and Kelvin can be done using the conversion equations between them that are shown in Table 1.8. The following script implements the conversion from Celsius to

TABLE 1.8:   Conversion formulae

Celsius to Fahrenheit	$°F = 9/5 \ (°C) + 32$
Kelvin to Fahrenheit	$°F = 9/5 \ (K - 273) + 32$
Fahrenheit to Celsius	$°C = 5/9 \ (°F - 32)$
Celsius to Kelvin	$K = °C + 273$
Kelvin to Celsius	$°C = K - 273$
Fahrenheit to Kelvin	$K = 5/9 \ (°F - 32) + 273$

Fahrenheit. As an example, let the Celsius temperature be 45 °C. The result is:

> Enter the temperature in Celsius degrees:
>
> 45
>
> The equivalent Faherenheit temperature is 113.00:

The script is:

```
'''Temperature conversion
This script converts between different temperature scales '''

# START

# Read the temperature in Celsius degrees:
C = float(input("Enter the temperature in Celsius degrees:\n"))

# Compute the conversion:
F = 1.8*C + 32

print("The equivalent Faherenheit temperature is %0.2f" %F)

# END
```

Example 1.8 Simple and compound interest

It is desired to compute the simple and compound interest paid to an investor at a fixed interest annual rate. The number of months n is divided by 12 to obtain the number of years. The formulae for simple and compound interest are:

> simple interest = principal $\times n \times$ interest rate
>
> compound interest = principal $\times$ (1 + interest rate^n) -1

To compute the power the function pow is used and it is available at the library math that must be imported. The script is:

```
"'Interest calculation.
This is Example 1.8. This program obtains the simple
and compound interest. '"


# Import required libraries:
import math


principal = float(input("Enter the principal:\n"))
interest_rate = float(input("Enter the interest rate: \n"))
n = float(input("Enter the number of months:\n"))


# Divide number of months by 12.
n = n/12


# Simple interest:
simple_interest = principal*n*interest_rate
compound_interest = principal*pow(1 + interest_rate, n) - 1


# Display results:
print("Simple interest: %0.2f" % simple_interest)

print("Compound interest: %0.2f" % compound_interest)
```

As an example consider a principal of $10,000.00, an annual interest rate of 4%, and 18 months (1.5 years). Running the program the results are:

```
Enter the principal:
10000
Enter the interest rate:
4
Enter the number of months:
18
Simple interest: 600.00
Compound interest: 605.96
```

The function **pow(a, x)** has the meaning:

$$\text{pow}(a, x) = a^x$$

TABLE 1.9:   Python instructions for Chapter 1

Instruction	Description
#	Comment line.
**	Power.
" "	String delimiter.
' '	String delimiter.
""" """	Comments delimiter.
&	Boolean AND.
|	Boolean OR.
and	AND instruction.
bool	Boolean variable.
False	False value of a boolean variable.
float	Floating-point type variable.
import	Import a module or library.
input	Read a variable.
int	Integer type variable.
len	Length of a string, list, etc.
list	List type variable.
math	Mathematics module or library.
None	None variable.
not	NOT instruction.
or	Boolean OR.
print	Display a variable.
pow	To obtain a number to the power of other number.
str	String variable.
True	True value of a boolean variable.
type	Display the type of a variable.

1.17   Python Instructions for Chapter 1

The Instructions used in Chapter 1 are displayed in Table 1.9.

1.18   Conclusions

In this chapter the basic Python concepts such as variables, scripts, modules or libraries, among many others, have been presented. Some examples show how easy is to have a running program in Python. The Python IDLE has been used but everything can be used in other platforms such as Jupyter or Spyder.

1.19 Exercises

1.1 What is displayed for the script presented below? Try to find the result without running the script in the IDLE.

```
n = 4; k = 2
print(n)
print(-n)
print(n-)
print(n + k)
print(k)
print(n, k)
print(" ")
print("n %3.2g" % n)
print("n * n = %3.2g" % n*n)
print(n*n)
print('n')
```

1.2 Indicate the output in the following segments of code:

a)
```
x = 3
x = x + 2
print(x)
```

b)
```
y = - 5.3
y = y*7.0
print(y)
```

c)
```
x = 8; y = 3)
x = x%y
print(x)
```

1.3 Find the results for the following segments of code:

a)
```
x = 3
x = x + 2
print(x)
```

b)
```
y = - 5.3
y = y*7.0
print(y)
```

c)
```
x = 8; y = 3
x = x % y;
print(x)
```

1.4 The function `getsizeof( )` from the library `sys` is used to obtain the size of a variable. Using `getsizeof( )` obtain the size of the variable "a" for

a) a = 2.0

b) a = 2

c) a = 20000

d) a = '2'

e) a = True

1.5 Indicate the result produced by the following code:

```
x = 42.0
y = 42.3 e-14
a = 42.3 e-14
b = -57.8
m = +87
n = +249
print(m*n)
print(x)
print(y)
print(a)
print(b)
```

1.6 Indicate the result produced by the following code:

a) x = 90.34
   print(x)

b) z = 2
   y = 7;
   x = y + 2*z
   print("The result is: %5.2g" % x)

1.7 Say if the following code produces `True` or `False`:

```
x = 8; y = 5
a = 1; b = 2
```

a) x = 1 & a <= 2 | y < x

b) (x = 1 & a < texttt= 2) | y > x

c) x = 1 & (a <= 2 | y = x)

1.8 What is the result in the following code?

```
a = 2; b = 3
x = 8.3; y = -4.6; z = 2.3
print ("a*b = %5.2f"  % a*b )
print("a/b = % 5.2f" % a/b )
print("x/y = % 5.2f" % x/y )
print("a*z = %5.2f" % a*z )
```

1.9 What is the result in the following code?

```
c = '8'
r = 3
print(c*r)
```

1.10 Obtain the result in every case:


```
x = 7; y = 3
```

a. x   % y

b. x = x + y

c. x += y

d. x =+ y

e. x *= y

f. x =* y

g. x %= y

1.11 Write a script to read the data: 7.0, 2.0, 8.3, 3.0, 5.6, 4.0 and compute:
   a. The maximum value.
   b. The minimum value.
   c. The addition.
   d. The mean.

Chapter 2

Conditionals and Loops

2.1 Introduction

In real life, very often it is necessary to take a decision about what is the next step in a given situation. The decision usually is taken depending upon certain conditions in the situation. In the Python programming language, this decision-making process also takes place, and the condition or conditions in the process have to be evaluated. The decision-making instructions are known as conditionals and they are studied in the first part of the chapter. The second part of the chapter studies the instructions to repeat a process. These instructions are called loops and they appear in iterative processes such as those in numerical analysis, physics and finance, to name a few areas.

The chapter starts with the definition of a conditional in Python. Then, the extension to nested conditionals is presented. Several examples show the use of conditionals. Loops are introduced next. The two types of loops, `for` loops and `while` loops are defined. Finally, examples using loops and conditionals are presented.

2.2 Conditionals

Conditionals arise when an operation execution depends upon the true value of a logic expression. For example, to take the square root of a number can only be done if the number is positive. Thus, a conditional has to be implemented to check that the number is positive.

The conditions in Python use the key word `if` in the following format:

```
if (logic expression):
        Instructions
Next instructions
```

The key word `if` is followed by a `logic expression` and a colon. The `Instructions` in the block must be indented by at least a single space but

DOI: 10.1201/9781003222118-2

they must have the same indentation. The indented instructions are part of
the if statement and they are executed if the logic expression is True. Otherwise, if the logic expression is False, the Next instructions are executed.
The logic expression can be inside parenthesis but this is not compulsory and
they may be omitted.

A very common error is to omit the semicolon.

Example 2.1 Simple conditional using if

To obtain the quotient of two numbers, a condition that must be satisfied is
that the denominator must be different from 0. This condition can be checked
with an if conditional. A script to evaluate the quotient of two numbers must
first read in the numerator and denominator and assign them to variables a
and b:

```
a = float(input("Enter numerator: "))
b = float(input("Enter denominator: "))
```

Next, initialize a variable check to False and with a conditional change it to
True if the denominator is not equal to 0:

```
check = False
if (b != 0):
    check = True
```

Now with other condition evaluate the quotient if check = False:

```
if (check):
    c = a/b
    print("Result:", c)
```

When the variable check is True, then it executes c = a/b and prints the
result. Otherwise, when a check is False it does not do anything.
The complete script is:

```
# This is script Example_2_1.py
# Logical Conditional.

a = float(input("Enter numerator: "))
b = float(input("Enter denominator: "))

# Script continues next page.
```

TABLE 2.1: Relational Operators.

Operator	Definition	Example	Traditional Notation
>	Greater than	A > B	A > B
>=	Greater than or equal to	A >= B	A $\geq$ B
<	Less than	C < 4	C < 4
<=	Less than or equal to	X <= 8	X $\leq$ 8
==	Equal to	X == Z	X = Z
!=	Different to	a != 8.2	a $\neq$ 8.2

```
# Script continued from previous page.

check = False
if  (b != 0):
    check = True
# End of conditional.


# Starts conditional.
if (check):
    c = a/b
    print("Result:", c)
```

The logic expression may use any of the relational operators shown in Table 2.1. It can be seen that the equality comparison uses a double-equal sign. Thus, a $==$ b is comparing if a is equal to b. If the equality holds then the logic expression is **True**. Otherwise, it is **False**. It is worth mentioning that a $=$ b is an assignment meaning that the value of b is assigned to a; thus, it is not a comparison.

To compare equality a double equal sign $==$ is required.

For example, consider the following example of correct and incorrect logic expressions:

if(a == b):	**CORRECT**	It is a comparison.
if(a = b):	**incorrect**	It is an assignment. It is an error.
if(a == b)	**incorrect**	The colon is missing :

TABLE 2.2:   Logic Connectives.

Operator	Definition	Example	Precedence
not	logical NOT	not X	1
&	logic AND	A & B	2
and		A and B	2
∧	exclusive OR	A ∧ B	3
|	logic OR	A | B	4
or		A or B	4

Composite logic expressions may also use the logic connectives from Table 2.2. Some examples are:

```
a = 1; b = 2; c = 3
if( a < b ):              True logic expression
if( a < b or c > 0):     True logic expression
if( b > 1 and c == 4):   False logic expression
```

Example 2.2 Solution of a first-degree equation
A first-degree equation is of the form $ax + b = 0$. The solution is given by $x = -b/a$, which requires that $a \neq 0$. Thus, a conditional is required to check this value. To solve the equation, the following steps are required:

- Read in a and b

- Check if a $\neq$ 0

- Solve for $x = b/a$

- Display x

In Python the code is:

```
# This is script Example_2_2.py
# Code to solve a first-degree equation.
# START:

# Enter the coefficient values:
a = float(input("Enter a: "))
b = float(input("Enter b: "))

# Script continues at next page.
```

```
# Script continued from previous page.

if (a != 0):
    x = - b/a
    print ("Result:", x)
# End of the conditional.
# END
```

When this script is executed for the equation $2x + 3 = 0$ the coefficient values are $a = 2$, $b = 3$ and the value for x is -1.5. For the equation $0x + 3 = 0$ the coefficient values are $a = 0$ and $b = 3$. In this case, the program does not give any answer because the conditional is not true and the solution is not obtained.

2.3   The Conditional `if-else`

In the previous examples, it might be desirable to print out a message when the division cannot be made. This can be made using a statement `if-else`. The format is:

```
if (logic expression):
        instructions 1
else:
        instructions 2
```

This statement works as before in the first part; that is, the `instructions 1` below the `if` conditional are executed if the `logic expression` is true, but in the second part, the `instructions 2` corresponding to `else` are executed when the `logic expression` is false. Note that there is a semicolon after the key word `else`.

The first-order equation script can be modified to print out a message as:

```
# This is script Example_2_2b.py
# Code to solve a first-degree equation.
# A statement if-else is used.
# START:

# Enter the coefficient values:
a = float(input("Enter a: "))
b = float(input("Enter b: "))

# Script continues at next page.
```

```
# Script continued from previous page.

if (a != 0):
    x = − b/a
    print ("Result:", x)
else:
    print ("The equation cannot be solved because a  = 0")
# End of the conditional.
# END
```

Note that the `else` instruction MUST be aligned with the `if` instruction.

2.4   Nested Conditionals

A conditional statement is nested if it is placed inside another conditional statement. Nested conditionals use the statement `if-else`. The format for a nested conditional is:

```
if (logic expression A):
    instructions 1.
else:
```

$$\text{Nested if} \begin{cases} \text{if (logic expression B):} \\ \quad \text{instructions 2.} \\ \text{else:} \\ \quad \text{instructions 3.} \end{cases}$$

```
    following instructions...
```

If `logic expression A` is True, then, `instructions 1` are executed and when finished, the script continues with the `following instructions....`

If `logic expression A` is False, then, in the nested conditional the `logic expression B` is checked. If it is True, `instructions 2` are executed and when finished the script continues with the `following instructions....` If the `logic expression B` is False, then the instructions 3 after the `else` are executed and then it continues executing the `following instructions....`

There is no limit as to how many conditionals can be nested but usually no more than three or four are used for the sake of clarity and maintenance. Comments should be used as much as possible to explain the process.

A common way to write the nested conditional is using the contraction for `else-if` written `elif`. The format for the nested conditionals is then:

```
if (logic expression A):
    instructions 1.
elif: logic expression B:
    instructions 2.
else:
    instructions 3.
following instructions....
```

Example 2.3 Solution of a second-degree equation

The second-degree equation provides an example that can be solved with Python using nested loops. A second-degree equation is of the form

$$ax^2 + bx + c = 0$$

Its solutions are given by

$$x_{1,2} = \frac{-b \pm \sqrt{b^2 - 4ac}}{2a} \tag{2.1}$$

The discriminant of this equation is given by

$$d = b^2 - 4ac \tag{2.2}$$

Depending upon the value of this discriminant there are three possible solutions as follows:

1. Check if $d > 0$

   - The solutions are real and different and given by equation (2.1).

2. Check if $d == 0$

   - The solutions are real and equal and given by:

$$x_1 = x_2 = \frac{-b}{2a}$$

3. Check if $d < 0$

   - The solutions are complex conjugate and they are given by:

$$x_{1,2} = \frac{-b \pm i\sqrt{4ac - b^2}}{2a}$$

where i is the imaginary number $i = \sqrt{-1}$.

The steps to solve the second-degree equation are then:

1. Read in the coefficients a, b y c.
2. Evaluate the discriminant $d = b^2 - 4ac$.
3. Depending upon the value of d evaluate x_1 and x_2.
4. Finally, display the values of x_1 and x_2.

To display the +- sign in Pyhton the unicode code \u00b1 is used as shown below. The implementation in a Python script is as follows:

```
# This is script Example_2_3.py

# START:

# Import the library math for the square root function:
from math import *

# Read in the coefficients:
a = float(input("Enter a: "))
b = float(input("Enter b: "))
c = float(input("Enter c: "))

# Evaluate the discriminant d:
d = b**2 - 4*a*c

# Start the conditional if-elif-else
# for the different values of d.
# Case of real and different roots.
if (d > 0):
    print("Roots real and different:")
    x1 = (-b + sqrt(d))/2/a
    x2 = (-b - sqrt(d))/2/a
    print('First root is equal to: ', x1)
    print('Second root is equal to: ', x2)

# Case of equal roots.
elif(d == 0): # Nested conditional.
    print("Roots real and equal:")
    x1 = -b/2/a
    x2 = x1
    print('Both roots are equal to: ', x1)

# Script continues next page
```

```
# Script continued from previous page

# Case of complex conjugate roots.
else:
    print('Roots complex conjugate:')
    x1_real_part = -b/2/a
    x1_im_part = sqrt(-d)/2/a
# To print the sign ± uses u'\u00b1' in Python.
    print(x1_real_part, u'\u00b1 i*', x1_im_part)
# End of conditional.

# END
```

When this script is executed for a = 1, b = 1, and c = 1, the result is:

```
Roots complex conjugate:
-0.5 ± i* 0.8660254037844386.
```

2.5   Exceptions and Errors

Exceptions arise when there is an operation that produces a numerical error, for example when a division by zero occurs. Exception handling is beneficial to the program because it makes the code more robust. Python exceptions are errors detected usually at run-time. If the code is not written to handle the exception then the program crashes and terminates abruptly. As an example, the following script:

```
a = 3
b = 0
c = a/b
print(b)
```

produces the following error:

```
Traceback (most recent call last):
  File "C:/exception1.py", line 2, in <module>
    c = a/b
ZeroDivisionError: division by zero
```

To avoid this `ZeroDivisionError`, the Exception instruction can be used. The format for an exception in the case of division by 0 is:

```
try:
   c = a/b
   print(c)
except ZeroDivisionError:
   print("Zero division error")
else:
   print("Succesful division")
```

The key words are **try**, **except**, and **else**. The key words **try** and **else** are followed by a semicolon. The original code is placed after the **try** key word. The error message is placed after the **except** key word. A semicolon is placed after the error message. The **else** is optional and can be used to display a success message. In the previous code, the **else** statement can be omitted.

It is not necessary to know the name of the error, just to know that there might be the possibility of a run-time error. Another way to write the exception is as follows:

```
a = 3
b = 0
try:
   c = a/b
   print(c)
except:
   print("Zero division error")
```

And the result is:

```
Zero division error
```

Another error might be to try to add an integer or a floating point number to a string which is not possible and, therefore, an error occurs. For example:

```
number = 3.1416
string = "Letter"
result = number + string
print(result)
```

The result is:

```
Traceback (most recent call last):
File "C:/noplus_string.py", line 5, in <module>
result = number + string
TypeError: unsupported operand type(s) for +: 'float' and 'str'
```

With an exception it becomes:

```
number = 3.1416
string = "Letter"
try:
        result = number + string
        print(result)
except:
        print("Types are different")
```

And the result is:

```
Types are different
```

Exceptions are useful whenever an instruction might not be able to be executed. Table 2.3 gives a list of exceptions.

2.6   Loops

In some algorithms, it is necessary to repeat a set of instructions several times. Instead of repeating the code, *loops* can be used. Loops are statements that repeat a set of instructions a given number of times. There are two types of loops: the `while` loop and the `for` loop. In the `while` loop, a set of instructions is repeated a number of times depending upon a `logic expression`. In the `for` loop, the set of instructions is repeated a given number of times. In this case, there is no `logic expression` involved.

In the `while` loop, the logic expression is checked just before the loop begins. In this way, it might be possible that the instructions inside the `while` loop are never executed if the condition is not met. In the `for` loop there is not a logic expression to be satisfied and the set of instructions inside the loop is repeated a fixed number of times. The combination of both the `while` and `for` loops together with the conditions of the previous chapter gives the designer tools to implement powerful and yet, simple and elegant algorithms.

2.7   The `while` Loop

A `while` loop repeats a set of instructions as long as a logic expression holds true. When the logic expression becomes false the loop ends.

TABLE 2.3: Most common exceptions

Exception	Description of error
AssertionError	Assert statement fails.
AttributeError	Attribute assignment or reference fails.
EOFError	`input()` function finds an end-of-file.
FloatingPointError	A floating point operation fails.
GeneratorExit	A generator's `close()` is called.
ImportError	Imported module is not found.
IndexError	Index of a sequence is out of range.
KeyError	A key in a dictionary is not found.
KeyboardInterrupt	User hits the interrupt key.
MemoryError	Operation runs out of memory.
NameError	Variable not found in the local or global scope.
NotImplementedError	Raised by abstract methods.
OSError	System operation causes a system-related error.
OverflowError	When the result of an arithmetic operation is too large to be represented.
ReferenceError	A weak reference proxy is used to access a garbage collected referent.
RuntimeError	Error does not fall under any other category.
StopIteration	Raised by the `next()` function to indicate that there is no further item to be returned.
SyntaxError	Raised by the parser when a syntax error is encountered.
IndentationError	Raised when there is an incorrect indentation.
TabError	When the indentation consists of inconsistent tabs and spaces.
SystemError	The interpreter detects internal error.
SystemExit	Raised by the `sys.exit()` function.
TypeError	When a function or operation is applied to an object of an incorrect type.
UnboundLocalError	When a reference is made to a local variable in a function or method, but no value has been bound to that variable.
ValueError	When a function gets an argument of correct type but improper value.
ZeroDivisionError	When the second operand of a division or module operation is zero.

The `while` has the following format:

```
while (logic expression):
      set of instructions
```

It can be seen that if the `logic expression` is false then the `while` loop may not be executed because the first thing that has to be done is to check if the `logic expression` is true.

The instructions inside the `while` loop must be **indented**. In addition, there must be present a variable called the *index* that must be initialized before the `while` loop and may change its value inside the loop to modify the `logic expression`.

It could be the case that the `logic expression` is always false and, therefore, the `while` loop NEVER executes.

Example 2.4 Sum of integers from 1 to 10

In this algorithm it is desired to implement

$$\text{sum} = 1 + 2 + 3 + 4 + 5 + 6 + 7 + 8 + 9 + 10$$

whose result is sum $= 55$.

To implement this sum a `while` loop is used. The algorithm is made for any end value. The script starts reading the number N where the `sum` ends, then it continues by **initializing** the `sum` and an index k to 0.

```
N = int(input("Enther the value of N: "))
k = 0
sum = 0
```

In the `while` loop the `logic expression` involves the index, so when it reaches the value of eleven the logic condition becomes false. Thus the `logic expression` is:

```
k < N + 1
```

The `sum` is incremented by the index, which takes the values from 1 to N. This is done with:

```
sum = sum + k
```

Inside the loop, the index k is incremented by 1 as k `+= 1`. The `while` loop is then

```
while k < 11:
      sum = sum + k
      k += 1
```

Finally, the sum is displayed as:

```
print("The sum is from 1 to N is equal to ", sum)
```

The complete script is

```
# This is script Example_2_4.py

# Initialization of index k and sum:

N = int(input("Enter the value of N: "))
k = 0
sum = 0

# The while loop starts:
while k < 11:
        sum = sum + k
        k += 1

# The result is displayed:
print("The sum is from 1 to 10 is equal to ", sum)
```

The result after running the script for N = 10 is 55 as expected. The variable where the sum is made is known as the **accumulator**.

Example 2.5 Looking for prime numbers

A number is prime if its only divisors are unity and itself. To check if a number N is a prime number, it can be divided by all the numbers less than N starting at 2 and ending in N - 1. If the residue of any of those divisions is = 0, it means that N is divisible by that divisor and, therefore, the number is not a prime number. It can be shown that it is only necessary to make divisions up to $\sqrt{N}$. An algorithm that implements this search using a while loop is:

1. Read in the integer N.

2. Initialize the divisor to 2.

3. Start the **while** loop if the divisor is less than $\sqrt{N}$. If this is the case implement quotient = N//divisor with integer division.

4. If N = divisor*quotient, then N is divisible by divisor and N is not a prime number. In this case, the loop is finished by changing the number N to 0.

5. If N $\neq$ divisor*quotient, the value of divisor is incremented by 1 as divisor + 1 and the **while** loop continues the search.

6. The **while** loop ends when a divisor for N is found or when the search ends because the `divisor` is greater than $\sqrt{N}$.

7. Once the `while` loop is finished, the result is displayed with a `print` instruction.

In this algorithm, the variable `divisor` is the index and it is incremented each time the `while` loop is executed.

The script in Python is:

```python
# This is script Example_2_5.py
# Import the function sqrt from the library math.
from math import sqrt

# Read in the integer N.
N = int(input("Enter the value for N: "))
# Initialize the divisor to 2.
divisor = 2

# Start the while loop:
while divisor < sqrt(N):
        quotient = N//divisor
        if N == divisor*quotient:
                # Not a prime.
                N = 0 # To exit the loop.

        # Not found yet.
        else:
                divisor += 1
if N == 0 :
        print("The number is NOT a prime")
else:
        print("The number is a prime")
```

The result for two numbers is:

```
Enter the value for N: 11
The number is a prime.

Enter the value for N: 8
The number is not a prime.
```

The following algorithm is a modification of the previous algorithm but it makes use of Boolean variables that can only take the values **True** or **False**.

Example 2.6 Another algorithm for prime numbers
An algorithm that uses a Boolean variable can be used to find out if a given number N is a prime number. A Boolean variable `is_prime` is introduced and is initialized to `True` and it is changed to `False` if the number is divisible by an integer less than the number N. The algorithm is shown as follows,

```
# This is script Example_2_6.py
# It checks if a given number is a prime.
# START
from math import sqrt
divisor = 2
is_prime = True
N = int(input('Enter the number: '))
while divisor <= sqrt(N) and is_prime:
    quotient = N//divisor
    if quotient*divisor == N:
        is_prime = False
    divisor = divisor + 1
if (is_prime):
    print("It is a prime")
else:
    print("It is not a prime")
```

As can be seen the comparison is made with the Boolean variable `is_prime`.

2.8 The `for` Loop

A `for` loop repeats a set of instructions a determined number of times and it must include a counter. The `for` loop has the format:

```
for counter_values:
        Instructions
```

In the `for` loop, it is not necessary neither to initialize the `counter_values` nor increment its value. These actions are automatically done by the `for` loop. The set of instructions inside the `for` loop are executed each time the `counter_values` is incremented up to the indicated final value. Note that the set of instructions is indented after the semicolon. The instructions inside the loop ends when the indentation is ended. Another way to express values for the `counter_values` is in a list as in

```
for counter in a list:

    Instructions
```

The following points have to be observed in a `for` loop:

- Before the `Instructions`, a semicolon must be written.

- **NO** parenthesis are used in the `for` loop line.

- The set of instructions in the **for** loop must be indented.

A couple of examples show the way to use the `for` loop.

Example 2.7 Sum of integers from 1 to 10
This is Example 2.4 but now using a `for` loop. The steps are the following:

- Initialize the sum.

- Use a **for** loop to compute the sum.

- The `counter` may be one the term in the addition.

The implementation in a Python script is:

```
# This is script Example_2_7.py
# Sum of integers from 1 to 10.
# This algorithm uses a for loop.
# START
# Initialize the sum and the list.
sum = 0
list = [1, 2, 3, 4, 5, 6, 7, 8, 9, 10]
for counter in list :
      sum = sum + counter
print ("The sum is: ", sum)
# END
```

The result of this program is:

    sum = 55.

Example 2.8 Number of vowels in a phrase

To find the number of vowels in a string containing a phrase, a `for` loop can be used. The number of vowels is stored in an integer variable called `count`. The procedure is to compare each character in the phrase to a base string containing the vowels, small and capital ones. This base is `base = "aeiouAEIOU"`.

    phrase = "The greyhound runs fast on the track"

The scriptwith a `for` loop is:

```
# This is script Example_2_8.py
# Program: Count the number of vowels in a phrase.
phrase = "The greyhound runs fast on the track"
base = "aeiouAEIOU"
count = 0

# For each character in the phrase.
for i in phrase:

    # Checks if the character is in the base.
    if i in base:
        count = count + 1
print("The number of vowels is ", count)
# END
```

The result after running the script is:

    The number of vowels is 9

2.8.1 The Function range

A function that is usually used in a `for` loop is the function **range**. This function generates a list. The format is:

    range(a, b, c)

This function creates a list that starts in a, it ends before b, and c is the increment. The list generated is:

    [a, a + c, a + 2c, a + 3c, ...,]

The value b is NOT part of the list. Some examples are:

range(5)	produces [0, 1, 2, 3, 4]
range(2, 6)	produces [2, 3, 4, 5]
range(-1, 7, 2)	produces [-1, 1, 3, 5]
range(3, 9, 2)	produces [3, 5 ,7]
range(8, 4, -1)	produces [8, 7, 6 , 5]
range(-8, -13, -1)	produces [-8, -9, -10 , -11, -12]

In every case, the last value in the list is NOT b. The last two examples show decrement. The value of the increment cannot be 0.

Example 2.7 can be improved with a function range(1, N+1). This is equivalent to a list [1, 2, 3, ..., N]. The last number N+1 is not included in the list. The algorithm can be changed to obtain the sum of the first N integers. The value of N has to be entered into the program. The script is changed to:

```
''' This is script Example_2_7b.py
 Sum of integers from 1 to 10.
 This algorithm uses a for loop.'''

# START

# Initialize the sum and read in N.
sum = 0
N = int(input("Enter the value of N: "))

# Start counting with a for loop
for counter in range(1, N) :
     sum = sum + counter

# Print out the result
print("The sum is: ", sum)

# END
```

2.9   Nested Loops

There are some algorithms that require the use of a nested loop; that is, a loop within another loop. The nested loop can be a for loop or a while loop. The for and while loops can be combined in the nesting. For example, a while loop can be nested inside a for loop and vice versa. The first loop is known as the external loop and the loops inside this loop are known as the nested loops. The structure of the nested loops is:

```
loop_1
    Instructions_1.

    # First nested loop.
    loop_2
        Instructions_2.

        # Second nested loop.
        loop_3
            Instructions_3.

        # End of loop_3.
    # End of loop_2.
# End of loop_1.
```

It can be seen that `loop_2` is nested inside `loop_1` and that `loop_3` is nested inside `loop_2`. Each nested loop must be completely inside another loop.

Example 2.9 Course grade average
In this example, a `while` loop is nested inside a `for` loop to read students grades for six months. The algorithm has to read the monthly grades for three students during a semester. The grades on a scale from 0 to 10 are:

Name	1	2	3	4	5	6
Gary	8.7	9	7.8	8.1	8.2	9.5
Laura	9.6	10	10	9.4	9.7	9.3
Alina	9.5	6.4	8.9	8.9	9.9	10

For each student, the algorithm must read the name of the student, and the grades, and calculate the average. The steps to follow are:

1. Read the name of the student. To count the number of students use a counter.

2. Read the grades, add them up, and take the average. Use a grade counter to count the number of grades that have been read.

3. Display the average.

4. Repeat the previous steps for the remaining students.

In the **for** loop a **range** function is used. The algorithm is:

```python
# This is script Example_2_9.py
# Algorithm: Grade average.

# Read in input data:
No_students = int(input("Enter the number of students:\n"))

# Initialize array and General average:

array = [ ]
GeneralAverage = 0

# The for loop starts.
for counterStudent in range(No_students):
# Read the name of a student:
    print("Enter the name of the student: ")
    NameStudent = input( )
    # Save the name to a list.
    list = [ ]
    # Initialize the sum for the student.
    sum = 0.0

    # Initialize the while loop index.
    grade_counter = 1
    # The nested while loop starts.
    # The while loop is repeated six times.
    while (grade_counter <= 6):
        # Read a grade:
        print("Enter a grade" )
        grade = float(input())
        # Append the grade to the list:
        list.append(grade)
        sum = sum + grade
        grade_counter += 1
    # End of while loop.

    # Average calculation.
    average = sum/6.0
    # Add the average to the list:
    list.append(average)
    # Update GeneralAverage:
    GeneralAverage += average

Script continues at next page.
```

```
# Script continued from previous page.

    # The list contains the student's name,
    # the grades, and the average.
    # Append the list to the array:
    array.append(list)
# End of for loop.

# Print the array that contains the names, grades,
    # and average using a for loop.
for row in array:
    print()
    for column in row:
        print('%s' %column, end = ' ')
print()
GeneralAverage /= No_students
print("The average for the class is: %0.2f" %GeneralAverage)
```

Running the script the result is:

```
Gary 8.7 9.0 7.8 8.1 8.2 9.5 8.549999999999999
Laura 9.6 10.0 10.0 9.4 9.7 9.3 9.666666666666666
Alina 9.5 6.4 8.9 8.9 9.9 10.0 8.933333333333334
The average for the class is: 9.05
```

Example 2.10 Algorithm to compute a function
It is desired to compute a function $f(x, y, z)$ for the following integer values
of x, y, and z:

Values for: $x = 0, 2, 4, 6$; $y = $ -3, -2, -1, 1, 2, 3; $z = 2, 5, 8, 11$.

where $f(x, y, z) = 3x^z - 7y^2 z + 4xyz$

The variables x and z have four values, and y has six values. Three loops have
to be used for the computation. Any combination of for and while loops can
be used. For the values for x a for loop is used, and a while loop for the y
and z values.
    The values for x can be written as the list x = [0, 2, 4, 6] and the y
and z values as the lists y = [-3, -2, -1, 1, 2, 3] and z = [2, 5, 8,
11]. They can also be written as:

```
for x in range(0, 7, 2):
for y in range(-3, 4, 2):
for z in range(2, 12, 3):
```

The algorithm needs to have the three loops nested. Thus, for the first value for x, say x0 = 0, and for the first value for y, say y0 = -3, compute the function for every value of (x0, y0, z). Then, use the following value for y as y1 = -1 and compute for every value (x0, y1, z). Continue in this way until the function is computed for all the values of (x0, y, z). Then, for the next value of x1, repeat the previous process until the function is computed for every value of (x1, y, z). Continue until every value of x is used.

The script with the nested loops is:

```
# This is script Example_2_10.py
# The first loop is the for loop for x:
for x in range(0, 7, 2):

    # Initialize the index for the while loop for y:
    y = -3
    while y < 4:

        # Initialize the index for the while loop for z:
        z = 2
        while z < 12:
            f = x + y + z
            # Print the values for x, y, z, and f.
            print('x = ', x, 'y = ', y, 'z = ', z, 'f = ', f)

            # increment z:
            z = z + 3

        # increment y
        y = y + 2
    print()
```

After running the script, a long listing for the values of the function $f(x, y, z) = x + y + z$ is produced. The last four lines are displayed here:

```
x = 6    y = 3    z = 2     f = 11
x = 6    y = 3    z = 5     f = 14
x = 6    y = 3    z = 8     f = 17
x = 6    y = 3    z = 11    f = 20
```

2.10 The Instruction break

The instruction **break** interrupts a loop; that is, it ends it. In the case of nested loops, this instruction ends the loop where it is located. An example follows.

Example 2.11 The Babylonian method to find the square root
The Babylonian method to find the square root of a number was devised circa 1500 BC. The algorithm is as follows:

1. It is desired to find the square root of N.

2. N_0 is the initial value for the square root.

3. Evaluate

$$N_{k+1} = \frac{1}{2}\left(N_k + \frac{N}{N_k}\right)$$

   where k = 0, 1, ..., 1000

4. A **for** loop is used. The loop is terminated when the difference between consecutive values of N_k is less than 1×10^{-6}. That is

$$|N_{k+1} - N_k| < 10^{-6}$$

5. After the criteria is satisfied, display the result and the error.

6. Evaluate and display the square root with the function sqrt from the module math. Display the difference between both results.

The algorithm is:

```
# This is script Example_2_11.py
from math import sqrt

# Initial value.
Nk = 1.0

# Read the value for N:
N = float(input('Enter N: \n' ))

# Script continues at next page.
```

```
Script continued from previous page.

# for loop starts.
for k in range(2, 100):
    # Evaluate the new value Nk1.
    Nk1 = (Nk + N/Nk)/2

    print("The new value for NK1 is: ", Nk1)
    # Print the difference between Nk and NK1.
    print("The error is: ", abs(Nk - Nk1))

    # Compare if the error is less than the 1e-6.
    if abs(Nk1 - Nk) <1e-6:
        break
    # Update the value of Nk.
    Nk = Nk1

# Display the results.
print( "The square root after  ", k-1, "iterations is ", Nk1)
print("The square root with sqrt is: ", sqrt(N))
print("The error is : ", abs(sqrt(N)- Nk1))
```

Trying the value N = 2702.5 the result after 10 iterations is (the last iteration is shown):

```
The square root after 10 iterations is 51.98557492228012
The square root with sqrt is: 51.98557492228012
The error is : 1.5949463971764999e-12
```

For N = 2 the result is:

```
The square root after 5 iterations is 1.414213562373095
The square root with sqrt is: 1.4142135623730951
The error is : 2.220446049250313e-16
```

It can be seen that the error between the function `sqrt` and Babylonian method is very small.

2.11   The Instruction continue

The instruction `continue` can be used inside a loop and it is associated with a conditional. The instructions placed immediately after the instruction `continue` are not executed and the iteration continues with a new value for the index. An example shows its use.

Example 2.12 Use of `continue`

It is required to design an algorithm that adds the even numbers divisible by 5. The numbers are in a sequence given by range(1, 100, 3) = 1, 4, 7, 10, ... The numbers must be then divisible by 2 and by 5. A **for** loop can be used as:

1. A variable for the `sum` must be initialized.

2. A **for** loop is used to generate the numbers.

3. A conditional checks if the number is divisible by 2 or divisible by 5. If these two conditions are met, the number must be added to the sum. Otherwise, the remaining instructions in the loop are skipped.

4. Finally, print the number and the sum.

5. The `continue` instruction must be used.

The algorithm in Python is implemented in the following script:

```
# This is script Example_2_12.py
# The algorithm adds even numbers less than 100
# and divisible by 5.
# In addition, the numbers must start at 1 and be
# incremented by 3. That is,
# the numbers are 1, 4, 7, ..., 97.

# Initialize the sum
sum = 0

# The for loop starts:
for x in range(1, 100, 3):
        if x%2 or x%5:
                continue
        print("Value of x = ", x)
        sum += x
        print("Value of sum = ", sum)
```

After executing the script the result is:

```
Value of x = 10
Value of sum = 10.0
Value of x = 40
Value of sum = 50.0
Value of x = 70
Value of sum = 120.0
```

It can be seen that only three integers less than 100 are divisible by 2 and by 5 simultaneously.

2.12　Additional Examples

This section presents additional examples using loops and nested loops.

Example 2.13 Factorial of an integer number

The factorial of an integer number n that is denoted by $n!$ is defined by

$$n! = 1 \times 2 \times 3 \times 4 \cdots \times n - 1 \times n$$

As an example, the factorial of 5 is given by:

```
>>> factorial_of_n = 1*2*3*4*5
>>> print("The factorial of 5 is: ", factorial_of_n)
```

and the result is

```
The factorial of 5 is: 120
```

Now, if the factorial of another number larger than 5 is needed, this previous result can be used. In the case that it is needed to obtain the factorial of 20, this can be done by multiplying the factorial of 19 times 20. The factorial of 19 can be obtained by multiplying the factorial of 18 times 19, and so on. By definition, the factorial of 0 and the factorial of 1 are both equal to 1. The following script obtains the factorial for any integer number:

```
# Read the integer number.
# Define the factorial of 1 as the initial value.
factorial = 1
# Using a for loop the factorial for the next integer
# can be computed as the factorial already computed times
# the following integer as:
for k in range(2, n + 1):
        factorial = factorial*k
# Print out the result:
print("The factorial of ", n, " is ", factorial)
```

The following script implements the algorithm:

```
# This is script Example_2_13.py
factorial_of_n = 1
print("Enter the integer number n: \n")
n = int(input( ))

# Start computation of factorial of n.
for k in range(2, n+1):
    factorial_of_n = factorial_of_n*k

print("The factorial of ", n, "is ", factorial_of_n )
```

Executing this script for n = 5 gives the expected result. The factorial for another integer can be obtained too. For instance, for n = 10 the result is 3628800. In Chapter 7 the factorial function is revisited when the concept of recursivity is presented. There is a factorial function in the math library and in the numpy math library. Then the factorial can also be computed as

```
>>> import math, numpy
>>> print(math.factorial(10))
>>> print(numpy.math.factorial(10))
```

In both cases the result is 3628800.

Example 2.14 Random number generation

To generate random numbers the library **random** must be imported. Another function for random number generation is available in the library numpy.random. To generate random numbers, the library **random** has to be imported first. Integer random numbers can be generated with the function randint. The format is randint(a, b) to generate an integer random number between a and b, then to generate more than a single random number a loop has to be used.

In every programming language, the so-called random numbers are in fact pseudorandom, this means that are generated by a function in a deterministic way. In the case of Python, the "Mersenne Twister" generator is used. This algorithm was developed by M. Matsumoto y T. Nishimura in 1998.

To generate random numbers the following steps have to be followed:

1. In a loop generate random numbers

2. In every cycle in the loop store the random number in a list.

The following Python script obtains ten random numbers between 10 and 20:

```
# This is script Example_2_14.py
# Imports the library random.randint
from random import randint

# Initialize the list.
num = [ ]
# Generates a random integer in each cycle of the loop.
for i in range(10):
    n = randint(0, 20)
    # It adds the number to the list num.
    num.append(n)

# It prints out the result.
print(num)
```

A posible result is:

```
[18, 16, 14, 18, 13, 21, 15, 15, 11, 16]
```

The results vary from computer to computer and from run to run due to the pseudorandom nature of the function `randint`.

Other functions to obtain random numbers are available in `numpy` which includes the `random` library. To get integer random numbers the function `numpy.random.randint` can be used. As an example, five random numbers between 1 and 23 can be obtained with

```
>>> numpy.random.randint(1, 23, size = 5)
array([10, 11, 2, 10, 20])
```

where 1 is the lowest number, 23 is the greatest number and five random numbers are going to be displayed in an `numpy` array, in this case, a one-dimensional array.

To obtain random floating numbers between 0 and 1, the function available is `numpy.random.rand`. For example to obtain random numbers in a 3×4 array use

```
>>> numpy.random.rand(3, 4)
array([[0.42680396, 0.1773794 , 0.9608085 , 0.70378251],
       [0.18838926, 0.82229386, 0.28981669, 0.33885616],
       [0.21026074, 0.34366188, 0.6523904 , 0.45511047]])
```

Random numbers with a normal distribution can be obtained with `numpy.random.randn(array size)`. For example,

```
>>> numpy.random.randn(3, 4)
array([[-0.44262916, -0.04642462, -1.37445931, 1.56985599],
       [-0.08164265, -0.29936052, -0.00566557, -0.60739981],
       [-1.49277666, 1.81893915, -0.16286686, -0.53728375]])
```

2.13 Python Instructions for Chapter 2

Table 2.4 shows the Python instructions used in Chapter 2.

2.14 Conclusions

In this chapter conditionals and loops have been covered. The use of these instructions allows changing the sequence of a program depending upon either the values of the variables or the computations needed. The conditionals can be implemented with `if` closedup, `if-else`, and `elif` instructions. The loops can be implemented with `for` and `while` loops. By means of examples, these

TABLE 2.4: Python instructions for Chapter 2

Instruction	Description
if	Keyword for a conditional.
==	Equality comparison symbol.
&	Logical AND symbol.
and	Keyword for logical AND.
!	Logical NOT symbol.
not	Keyword for logical NOT.
∧	Logical exclusive OR symbol.
==	Equality comparison symbol.
|	Logical OR symbol.
or	Keyword for logical OR.
else	Part of a conditional.
elif	Keyword for if-else.
try	Keyword for an exception.
except	Keyword for an exception.
randint	Generates integer random numbers.
range	Generates a list.
continue	It is used to jump instructions in a loop.
brake	It is used to interrupt a loop.
for	It is used to create a loop.
while	It is used to create a loop.
random	Library to generate random numbers.

instructions have been presented. Another kind of conditional is the exceptions that can be used in case an error might arise in the program. Nested conditionals and nested loops have also been studied. All of these instructions can be used to implement more complex Python programs.

2.15 Exercises

1. What has to be done to an assignment in Python located immediately after an `if` instruction to indicate that the assignment must be executed only when the condition is true?

   (a) Underline the condition.

   (b) Start the instruction with a character "#".

   (c) Indent the assignment after the `if`.

   (d) Begin the assignment with a brace {.

2. Which one of the following operators is not a comparison one?

   (a) >=

   (b) !=

   (c) <

   (d) =

   (e) ==

3. When there are multiple lines in an `if` block, how is the end of the block indicated?

   (a) The same indentation must be applied to the lines in the block.

   (b) Use braces to indicate the block corresponding to the `if` instruction.

   (c) Capitalize the first word not belonging to the `if` block.

4. What is the Python reserved word used when the logical test fails?

   (a) `switch`

   (b) `else`

   (c) `otherwise`

   (d) `break`

5. What is the result in the following code?

```
x = 0
if x < 2:
    print('Small')
elif x <10 :
    print("Medium')
else:
    print('Large')
print ('FINISHED')
```

   (a) Large
      FINISHED
   (b) Small
      FINISHED
   (c) Medium
      FINISHED

6. For the following code::

```
if x < 2 :
    print('Less than 2')
elif x >= 10:
    print("Two or more")
else:
    print('Something else')
```

   What value of x will display 'Something else'.

   (a) x = 2
   (b) x = -2
   (c) This code will never display 'Something else' regardless of the value of x. .
   (d) x = -2.0

7. Which of the following statements are `True` or `False`?

```
a = 1; b = 2; c = 3
x = True
```

Logic expression	Result
a < b	
a < c & b > c	
a < c & x	
b == a ∣ a < c	

8. A vending machine sells candies at $2.00 a piece. Write a program that when the customer pays with a $5.00 or a $10.00 coin gives the correct change and gives the instruction to the machine to dispatch a candy.

9. What is the result in the following loop?

```
n = 7
while n > 0:
        print(n)
print("Job finished.")
```

   (a) It is an infinite loop.

   (b) The instruction `print` must be indented by two blank spaces.

   (c) `while` is not a reserved word in Python.

10. What is displayed in the following Python script?

```
total = 0
for i in [5, 4, 3, 2, 1] :
        total = total + 1
print(total)
```

   (a) 5

   (b) 1.00

   (c) 10

   (d) 15

11. What is the iteration variable in the following script?

```
friends = ["Louie", "Dewey", "Huey"]
for friend in friends:
        print("Hello my friend ", friend)
print("Finished")
```

(a) friend

(b) friends

(c) Huey

(d) Dewey

(e) Louie

12. What is displayed by the following code?

```
The_smaller = -1
for number in [9, 12, 3, 9, 17, 23] :
        if number < The_smaller:
                The_smaller = number
print(The_smaller)
```

(a) -1

(b) 3

(c) 23

(d) 9

13. How many time is the following code executed?

```
n = 0
while n > 0 :
        print (n)
        n = n + 1
print(" Finished ")
```

14. What is the result in the following code?

```
for num in range(0, 20, 2):
    print(num)
```

15. What is the result in the following Python code?

```
for num in range(20, 0, -1):
    if (valor%4 != 0):
        print(num),"\n"
```

16. Use a `while` in the previous exercise.

17. Develop an algorithm and its corresponding Python script to generate 20 random integer numbers between 0 and 30. If the random number has been generated already, discard it and generate another number. Display the list of random numbers and how many times the loop was executed. Use a `for` loop.

18. Write a script that reads the names of persons. Display the names and end when a name is repeated. Use the comparison ==.

19. Write an algorithm that recursively computes the value of the constant *e* from the recursive equation

$$x_n = \left(1 + \frac{1}{n}\right)^n$$

To check how fast is the convergence, display the result in the 5th, 10th and 15th iteration.

2.16 Bibliography

1. https://docs.python.org/3/library/exceptions.html#bltin-exceptions

2. M. Matsumoto y T. Nishimura, Mersenne twister: a 623-dimensionally equidistributed uniform pseudo-random number generator, ACM Trans. Modeling and Computer Simulation, vol. 8, No. 1, pp. 3-30.

Chapter 3

Data Structures: Strings, Lists, Tuples, and Dictionaries

3.1 Introduction

A brief introduction to some data structures is given. The structures covered are strings, lists, dictionaries, and tuples. In this chapter, some functions that are applied to data structures are presented. Examples are used for this purpose.

3.2 Strings

Strings are data structures with alphanumeric elements. The elements of a string can be a single character, numbers, or a sequence of characters. The elements of a string are enclosed in quotation marks. Examples of strings are:

a101 = "Lucky Puppy 101"   car = 'Torino'   animal = 'horse'

In each one of the examples shown, the strings have a name and a certain number of characters. The string "a101" has 15 characters including the blank spaces, "car" has 6 characters, and "animal" has five. A string has to be delimited by either single or double quotes.

Each character has a position in the string. The leftmost character has position 0, the next one is position 1, and so on. Another way to indicate the position is the following: the right most character is in position -1, the following character to the left is in position -2, and so on. Thus, in string "car", the character "T' is in position 0, the character o is in position 1, etc. At the same time, the last "o" in "Torino" is in position -1, the "n" is in position -2, and it can be seen that the "T" is in position -6.

The position of a character in a string is an integer number known as the index.

DOI: 10.1201/9781003222118-3

The index indicates the position of a character in the string.

Sometimes it is necessary to select a character of the string. This can be done with the name of the string followed by the position enclosed in brackets. For example, for the string game = "base-ball", the letter e has the position 3 and then is game[3].

A segment of the string as game[5:8] = "bal" is called a slice. The notation [5:8] indicates the character from character 5 to character 8. Note that 'game[8] = l' is not included in the answer. Using the alternate notation, this same slice can be written as game[-4:-1].

---

x[k : n+1] selects the elements from k-th to the n-th.

Example 3.1 Slice of a string
Consider the string

    planets = "Mars_Jupiter_Saturn"

has 19 characters numbered from 0 to 18. The slice for "Jupiter" is obtained as planets[5:12], and the slice for Mars is planets[0:4]. This is obtained as:

```
>>> planets[5:12]
'Jupiter'
>>> planets[0:4]
'Mars'
```

To select the last planet, the following can be used

```
>>> planets[13:30]
'Saturn'
```

In this last slice, Python only prints the slice from character 13 to character 18 because the remaining characters do not exist.

For a string a, the operator a[: k] takes the slice that starts at the 0 position and ends at the (k - 1)-eth position, whereas the operator a[k:] is a slice consisting of the characters from the k-th position to the last one.

| x[k :] | Slice from index k up to the end of the string. |
| x[: k+1] | Slice from the beginning of the string up to k position. |

Two operations on strings are available, they are the sum of strings, also called concatenation, and multiplication by an integer.

3.2.1   Concatenation or Sum and Multiplication of Strings

The result of the concatenation or sum of two or more strings is another string. Concatenation can be made using the operator + between the strings.

Example 3.2 Concatenation of strings
Given the strings:

a = "The dog"   b = "is white"   c = "and black"

if the concatenation is made:

```
>>> new = a + b
"The dogis white"
```

There is not a blank space between the two strings because they only were concatenated. A blank space can be added to correct as:

```
>>> new = a + " " + b
"The dog is white"
```

Next, the three strings are concatenated and blank spaces are added between strings and a final period is also added:

```
>>> complete = a + " " + b + " " + c + "."
"The dog is white and black. "
```

The multiplication of a string by an integer number repeats the string as many times as the integer number indicates.

Example 3.3 Multiplication of a string by an integer
For the string:

triple = "Sun Earth Moon "

The result when it is multiplied by 2 is:

```
>>> print(2*factor)
```

"Sun Earth Moon Sun Earth Moon "

If the numeric factor is not an integer, Python sends an error message.

3.3 Functions on Strings

There is a number of functions that can be applied to strings. In this section some of the functions are presented.

3.3.1 Length of a String

The instruction `len` produces the length of the string; that is, the number of characters in the string including blank spaces, if any.

Example 3.4 Length of a string
It is desired to compute the length of the string "A picture is worth a thousand words". This can be done with:

```
>>> len( "A picture is worth a thousand words")
```

The result is 35, counting the blank spaces as parts of the phrase:

3.3.2 Split of a String

Sometimes it is necessary to split a large string, for example, to separate the words in it. The instruction `split` can be used in this case. The format is:

```
string_name.split(argument)
```

The argument in the split parenthesis means that the `string_name` is separated in that argument. If there is no argument then the string is separated into the blank spaces.

Example 3.5 Splitting a string
Carl Sagan's phrase in the string:

```
Sagan = "We live in a society exquisitely dependent on science
    and technology, in which hardly anyone knows anything about
    science and technology. "
```

First, it is desired to separate the phrase in the blank spaces; thus, the argument of the split function is blank:

```
>>> Sagan.split( )
```

['We', 'live', 'in', 'a', 'society', 'exquisitely', 'dependent', 'on', 'science', 'and', 'technology,', 'in', 'which', 'hardly', 'anyone', 'knows', 'anything', 'about', 'science', 'and', 'technology.']

But if the argument of the split function is "te", then:

```
>>>  Sagan.split('te')
```

['We live in a society exquisi', 'ly dependent on science and ', 'chnology, in which hardly anyone knows anything about science and ', 'chnology.']

It can be seen that the string separator 'te' does not appear in the result.

3.4   Immutability of Strings

Strings are immutable. This means that once a string is defined none of the characters can be changed, replaced or omitted. If a character or a set of characters has to change then a new string is defined.

Example 3.6 Immutability of strings
The following string is written in small letters:

```
city = 'puebla'
```

If it is desired to change the 'p' by 'P', the following could be done:

```
>>> city[0] = 'P'
```

but since strings are immutable, Python will give an error message:

```
Traceback (most recent call last):
File "<pyshell#60>", line 1, in <module>
>>> city[0] = 'P'
TypeError: 'str' object does not support item assignment
```

To overcome this error, a new string can be defined as:

```
>>> city2 = 'P' + city[1:6]
'Puebla'
```

It is possible to assign the same name to the new string:

```
>>> city = 'P' + city[1:6]
'Puebla'
```

3.4.1 Functions on Strings

There are several functions that operate on strings. Table 3.1 presents a list of functions that operate on strings. The following example shows several of the functions in the table.

Example 3.7 Functions on strings
For the strings:

a = "animals"

b = "flowers"

c = "nations"

The following functions can be applied:

Capitalize the first letter in the string:

```
>>> a.capitalize()
'Animals'
```

Place 'p' before and after the string the substring indicated to have a total of 20 characters:

```
>>> a.center(20,'p')
'pppppppanimalspppppp'
```

How many 'n' there are between the characters in positions 0 and 6:

```
>>> c.count('n', 0, 6)
2
```

Return True if the string ends with "ls":

TABLE 3.1: Functions on strings.

Method	Description
capitalize()	Converts the first character to upper case.
casefold()	Converts string into lower case.
center()	Returns a centered string.
count()	Returns the number of times a specified value occurs.
encode()	Returns an encoded version of the string.
endswith()	Returns True if the string ends with the specified value.
expandtabs()	Sets the tab size of the string.
find()	Searches the string for a value and returns the position.
format()	Formats specified values in a string.
format_map()	Formats specified values in a string.
index()	Searches the string for a value and returns the position.
isalnum()	Returns True if all characters are alphanumeric.
isalpha()	Returns True if all characters are in the alphabet.
isascii()	Returns True if all characters are ascii characters.
isdecimal()	Returns True if all characters are decimals.
isdigit()	Returns True if all characters are digits.
isidentifier()	Returns True if the string is an identifier.
islower()	Returns True if all characters are lower case.
isnumeric()	Returns True if all characters are numeric.
isprintable()	Returns True if all characters are printable.
isspace()	Returns True if all characters are whitespaces.
istitle()	Returns True if the string follows the rules of a title.
isupper()	Returns True if all characters are upper case.
join()	Joins the elements of an iterable to the end of the string.
ljust()	Returns a left justified version of the string.
lower()	Converts a string into lower case.
lstrip()	Returns a left trim version of the string.
max()	Returns the character with highest numeric value.
min()	Returns the character with lowest numeric value.
maketrans()	Returns a translation table to be used in translations.
partition()	Returns a tuple where the string is partitioned into three parts.
replace()	Returns a new string where a value is replaced.
rfind()	Searches the string for a value and returns the last position.
rindex()	Searches the string for a value and returns the last position.
rjust()	Returns a right justified version of the string.
rpartition()	Returns a tuple where the string is partitioned into three parts.
rsplit()	Splits the string at the separator and returns a list.
rstrip()	Returns a right trim version of the string.
split()	Splits the string at the specified separator and returns a list.
splitlines()	Splits the string at line breaks and returns a list.
startswith()	Returns True if the string starts with the value.
strip()	Returns a trimmed version of the string.
swapcase()	Swaps cases, lower case becomes upper case and vice versa.
title()	Converts the first character of each word to upper case.
translate()	Returns a translated string.
upper()	Converts a string into upper case.
zfill()	Fills with the specified number of 0 values at the beginning.

```
>>> a.endswith('ls')
True
```

Return **True** if all characters are alphanumeric:

```
>>> a.isalnum( )
True
```

Return **True** if all characters are numeric:

```
>>> a.isnumeric( )
False
```

min(string) returns the character with the lowest numeric value:

```
>>> min(c)
'a'
```

max(string) returns the character with the highest numeric value:

```
>>> max(c)
't'
```

Replace the character "s" with character "S":

```
>>> c.replace("s", "S")
'nationS'
```

Fill with zeros at the beginning of the string for a total of 10 characters:

```
>>> c.zfill(10)
'000nations'
```

Join with a given string a list of strings. In the example, the list
['dog', 'cat', 'bird'] is called the iterable:

```
>>> ' 123 '.join(['dog', 'cat', 'bird'])
'dog 123 cat 123 bird'
```

Given a substring, partition a string into a tuple consisting of the slice before the substring, the substring, and the slice after the substring:

```
>>> 'print('All roads lead to Rome'.partition('lead'))
'(('All roads', 'lead', 'to Rome')
```

3.4.2   Conditions and Loops

Conditions can be used on strings to check if characters belong to the string. For example, for the string given by school = 'University': :

```
>>> "u" in school
False
```

The answer is False because the character "u" is not in the variable 'school'. However, the character U produces:

```
>>> "U" in school
True
```

Example 3.8 Condition on a string

Consider the string: phrase = "May the best team win". Every time that the character "e" appears in the string, the string "player" must be displayed. If the character is not in the string then display the string "No game". Then:

```
if 'e' in phrase:
      print('Player' )
```

The result is:

```
Player
```

because the character "e" is found in the string "May the best team win". The string player is displayed only once because the first time the "e" appears in the string the condition is terminated. In the case that it is desired to print the string "Player" each time that character "e" appears in the phrase, a for loop can be used as:

```
phrase = 'May the best team win'
for i in phrase:
      if i == 'e':
            print('Player')
```

The result is now:

```
Player
Player
Player
```

because the character 'e' appears three times in the string phrase.

3.5 Lists

A list is a set of variables of the same or different type. A list has to be enclosed by square brackets and the variables, called elements, have to be separated by commas. Examples of lists are:

    birds = ['sparrow', 'pigeon', 'roadrunner', 'eagle']

    physics_constants = [9.81, 'Planck's constant', 'gravity']

    Irrational_numbers = [π, e, $\sqrt{2}$]

    b = ["car", "bike", 'scooter', 14, -3]

    c = [27,]

The first list is a list of strings and the fourth list has strings and numbers.

The last list is defined by a single element. It may include a comma but this is not necessary.

Each element of a list has associated an index to indicate its position in the list. The leftmost element has an index $= 0$, the next one to the right has an index 1, and so on. Another way to enumerate the index is starting with the rightmost element that is assigned index -1, the next one to the left has index $= -2$, and so on.

3.5.1 Operations on Lists

The operations that can be made with lists are sum and multiplication in the same way they were defined for strings.

Example 3.9 Length of a list
Obtain the length of the list `countries`:

    countries = ["Mexico", "France", "Japan", "Australia"]

    >>> len(countries)
    4

This answer is 4 because the list has four countries, but

    >>> len(countries[0])

6

because now the element 0 in the list of `countries` is the string "Mexico" that has 6 characters.

Example 3.10 Slices of a list
For the list:

```
numbers = [23, 45, 67, 89.4, -31]
```

a slice from the element 1 to the element 3 is a list from the second to the fourth elements, is:

```
>>> numbers[ 1:4 ]
[45, 67, 89.4]
```

and it can be seen that element with index $= 4$ is not included in the result.

3.5.2   Mutability

Lists are mutable. This means that the elements can either be changed, deleted, or have new elements added.

Example 3.11 Mutability of lists
For the list:

```
grades = [ 8.3, 9.45, 10, 9.9, 7.8 ]
```

If the grade 7.8 is to be changed to 8.5, it can be done with:

```
>>> grades[4] = 8.5
>>> print(grades)
[8.3, 9.45, 10, 9.9, 8.5]
```

The desired element changed its value because lists are mutable.

3.5.3   Conversion from a List of Strings to a Single String

It is possible to convert a list of strings to a single string with the instruction `join`. The substring that separates the individual strings in the list must be specified. The format is:

```
substring.join([list of strings])
```

Example 3.12 Functions on lists

For the list of strings and the substring:

```
>>> list = ["dogs", "cats", "mice"]

>>> substring = [" and "] # There is a blank space before
# and after and.
```

The elements in the list are joined by:

```
>>> substring.join(list)
```

To obtain:

```
'cats and dogs and mice'
```

If the substring is empty:

```
substring = ""
```

It produces:

```
>>> substring.join(list)
'catsdogsmice'
```

Example 3.13 Additional examples on lists

For the following lists:

The following functions are applied:


```
# Obtain the length :
>>> len(a)
5

# Obtain the maximum with the function max
```

```
>>> max(b)
1858

# Obtain the minimum with the function min:
>>> min(b)
-2702.2

# Add a new element to the list with the function append:
>>> a.append(-2015)
>>> a
['Peter', 555, 'Python', 'Intel', 'HP', -2015]

# Extend the list a with the list c:
>>> a.extend(c)
>>> a
['Peter', 555, 'Python', 'Intel', 'HP', -2015, 'mouse', 'keyboard']

# Insert 47 at position 2 of the list b:
>>> b.insert(2, 47)
>>> b
b = [45, -67.8, 47, 999.99, -2702.2, 1858]

# Return and remove the last element of list b:
>>> b.pop()
1858

# Return and remove the element at position 3 of the list b:
>>> b.pop(3)
999.99


# Remove the last element at position 3 of the list b:
>>> b.remove(45)
 [-67.8, 999.99, -2702.2, 1858]

# Reverse the order of the elements of the list:
>>> a.reverse()
>>> a
['keyboard', 'mouse', -2015, 'HP', 'Intel', 'Python', 555, 'Peter']

# Sort the elements of the list:
>>> b.sort()
>>> b
[-2702.2, -67.8, 45, 999.99, 1858]
```

3.6   Tuples

A tuple is a data structure consisting of elements that can be of different types. Tuples are IMMUTABLE which means that the elements of a tuple cannot be changed. Tuples may be enclosed between parentheses but that is not compulsory. Examples of tuples are:

```
champions = ( "Germany", "France", "Brazil", "Italy" )

numbers = ( 1, 233, 1.37e4, -456 )
```

3.6.1   Tuple Assignment

Tuple assignment provides a way to, for example, swap variables. This is so because in tuple assignment all the tuple elements are evaluated before the assignments.

Example 3.14 Swaping of variables
It is desired to swap the values of a = 23, b = -39 and c = 'car'. This is easily done with tuples as:

```
>>> (a, b) = (b, a)
>>> print(a, b)
```

The result is:

```
(-39, 23)
```

That is, the values in a and b have been swapped. This can be done with any number of elements in the tuple, for example:

```
>>> a, b, c = c, a, b
>>> print(' a = ', a, '\n b = ', b,'\n c = ', c)
```

to obtain

```
a = car
b = 23
c = -39
```

TABLE 3.2: Functions on tuples.

Funtion	Description
tuple(list)	Converts a list to tuple.
len(tuple)	Length of a tuple.
max(tuple)	Returns the element with the maximum value.
min(tuple)	Returns the element with the minimum value.

3.6.2 Functions on Tuples

Some functions available for tuples are shown in Table 3.2. Some examples show their use.

Example 3.15 Examples with tuples
For the following tuples:

```
a = ( -35, 4.67, 27.41, -98.65)
b = ( -35, 4.67, 27.41, 98.65)
```

The following functions can be applied:

```
# Convert tuple to a list
>>> c = list(b)
b = [ -35, 4.67, 27.41, 98.65]

# Obtain the length of a tuple
>>> len(b)
4

# Obtain the maximum of a tuple
>>> max(b)
98.65

# Obtain the minimum of a tuple
>>> min(b)
-35
```

3.7 Dictionaries

A dictionary is a data structure consisting of pairs of data. Each pair has a first element called the **key** and a second element called the **value**. Both the key and the value can be of any type. Dictionaries are delimited by braces. The format of a dictionary is:

```
a = { "key1":value1,"key2":value2,...,"keyM":valueM }
```

An example of a dictionary is:

```
grades = { "Hugo": 9,'Sarah': 9.2,"Bob": 7.8,"Xavier":10}
```

Dictionaries are not immutable; that is, the elements of a dictionary can be changed.

Pairs in dictionaries are referenced by the key; thus the pairs may appear in different order with respect to the way they were referenced. For example,

```
>>> print(grades)
{ 'Bob'  : 7.8, 'Xavier'  : 10, 'Hugo'  : 9, 'Sarah'  : 9.2 }
```

Only the key is needed to select a value:

```
>>> grades[ 'Bob' ]
7.8
```

Given that dictionaries are mutable, it is possible to change a grade, for example, to change Bob's grade from 7.8 to 8.9:

```
>>> grades[ 'Bob' ] = 8.9
```

Printing the grades again:

```
>>> print(grades)
{ 'Bob'  : 8.9, 'Xavier'  : 10, 'Hugo'  : 9, 'Sarah'  : 9.2 }
```

now Bob has the new value.

3.7.1   Functions for Dictionaries

Functions that can be applied to dictionaries are shown in Table 3.3. An example shows the different functions of this table.

Example 3.16 Example with dictionaries
The following functions can be applied to the following dictionary:

```
fruits = { "A" : "apple", "B" : "banana", "L" : "lemon" }

# Convert dictionary to string:
```

TABLE 3.3: Functions on dictionaries.

	Function	Description
1	str(dictionary)	Converts the dictionary in string.
2	type(variable)	Returns the type of variable.
3	dict.clear( )	Clears the dictionary.
4	dict.copy()	Copies dictionary in other variable.
5	dict.fromkeys(list,value)	Creates a new dictionary.
6	dict.get(key)	Returns the key.
7	dict.has_key(key)	Returns True if the key exists.
8	dict.items()	Returns a list of pairs (key, value).
9	dict.keys()	Returns a list with the keys.
10	dict1.update(dict2)	Add pairs key-value from dict2 to dict1.
11	dict.values( )	Returns a list with the values.

```
>>> myString = str(fruits)
myString = " { 'A' : 'apple', 'L' : 'lemon', 'B' : 'banana' } "

# Returns the type:
>>>  type(fruits)
< class 'dict' >

# Copies the dictionary into another variable
>>> fruits2 = fruits.copy( )
fruits2 = { "A" : "apple", "B" : "banana", "L" : "lemon" }

# Create a new dictionary:
>>> three = fruits.fromkeys( 'G' , 'grapes' )
three = { 'G' : 'grapes' }

# Returns a list of pairs:
>>> fruits.items( 'U': 'grapes')
dict_items ( [ ( 'A', 'apple'), ('L', 'lemon'), ( 'B', 'banana' ) ] )

# Returns a list with the values:
>>> three = fruits.values( )
dict_values( [ 'apple', 'lemon', "banana" ] )
```

TABLE 3.4:   Functions on sets.

	Function	Description
1	set(list)	Converts the list in a set.
2	type(set)	Returns the type of variable.
3	set.clear( )	Clears the set.
4	set.copy()	Copies the set in other variable.
5	set()	Creates an empty set.
6	set.add(element)	Adds an element to the set.
7	set.remove(element)	Removes an element from the set.
8	set.discard(element)	Removes an element from the set.
9	set.pop()	Removes the first element from the set.
10	set(string)	Creates a set with the characters in the string.
11	s1.union(s2)	Union of sets s1 and s2.
12	s1	s2)	Union of sets s1 and s2.
13	s1.intersection(s2)	Intersection of sets s1 and s2.
14	s1 & s2	Intersection of sets s1 and s2.
15	s1.difference(s2)	Set with elements in s1 but not in s2.
16	s1 - s2	Set with elements in s1 but not in s2.
17	s1.symmetric_difference(s2)	Set with elements in either one but not in both.
18	s1 ∧ s2	Set with elements in either one but not in both.

3.8   Sets

Sets are unordered collections of objects which do not allow repetition of the elements of the set. Any duplicate element is ignored. Sets support mathematical operations like union, intersection, difference, and symmetric difference. A collection of elements is a set if its enclosed by curly brackets. Examples of sets are:

```
fruits = {'watermelon', 'cherry', 'orange', 'apple'}
numbers = {21, 10, -17, 44, -37, 0}
mix = {0, 'apple', ''hammer', 34, 0, '0'}
```

Example 3.17 Example with sets
Table 3.4 gives some functions that can be applied to sets. For the sets `fruits`, `numbers`, and `mix` given above, and the list a = [3, 'car', 'dog', 27] the following can be computed:

```
# Convert list to a set:
>>> b = set(a)
```

```
b = {3, 'car', 'dog', 27}

# Returns the type:
>>> type(fruits)
<class 'set' >

# Copies the set into another variable
>>> fruits2 = fruits.copy( )
fruits2 = { 'watermelon', 'cherry', 'orange', 'apple'}

# Clears a set:
>>> fruits2 = fruits.clear( )
fruits2 = set( )

# Copy a set:
>>> fruits2.copy(fruits)
fruits2 = { 'watermelon', 'cherry', 'orange', 'apple'}

# Add an element to the set:
>>> fruits3 = fruits.add('pineapple')
fruits3 = { 'watermelon', 'cherry', 'orange', 'apple', 'pineapple'}

# Remove an element from the set:
>>> fruits4 = fruits.remove('apple')
fruits3 = { 'watermelon', 'cherry', 'orange', 'pineapple'}

# Discard an element from the set:
>>> fruits5 = fruits.discard('cherry')
fruits = { 'apple', 'watermelon', 'orange', '}
```

3.8.1   Sets by Comprehension

Sets can also be defined by comprehension. The format is

```
a = {x for x in function if condition}
```

Example 3.18 Generate a set by comprehension
Generate by comprehension a set for the even integer numbers between 1 and
10. This can be done with

```
>>> even_numbers = {x for x in range(11) if x//2*2 = x}
numbers = {0, 2, 4, 6, 8, 10}
```

3.8.2 Frozen Sets

The sets seen previously are mutable; that is, they can be modified by adding or deleting elements. Another type of sets which are immutable are frozen sets. They can be defined with the keyword **frozenset** and they cannot be modified. They can only accept an argument that can be any type of data structure. For example, for the set

```
>>> a = frozenset([2, 'dog', 'cats'])
```

If now an element is added:

```
>>> a.add(99)
```

The result is

```
Traceback (most recent call last):
  File "<pyshell# 29>", line 1, in <module>
    a.add(99)
AttributeError: 'frozenset' object has no attribute 'add'
```

Thus, most of the functions or methods in Table 3.4 cannot be used because of the immutability property of frozen sets.

3.9   Python Instructions for Chapter 3

The instructions used in Chapter 3 are summarized in the Table 3.5.

TABLE 3.5: Instructions for Chapter 3

Instruction	Description
frozenset	Defines a set as immutable.
x[n: m + 1]	Selects characters from the string x from n to m.
x[: m + 1]	Selects characters from the start to m.
x[n :]	Selects characters from n to the end.
len(x)	Number of characters in the string x.
split	String separator.
+	Concatenates strings.
*	Multiplies a string by an integer.
join	Join strings.

3.10   Conclusions

In this chapter strings, lists, tuples and dictionaries were defined as well as some properties related to these data structures. The exercises presented showed the different operations and functions on them. The examples at the end of the chapter provide a complement to learn to use these data structures.

3.11   Exercises

1. Concatenate the string "`Continent`" with the string "`American`" to obtain the string: "`American Continent`". There must be a blank space between the two words.

2. Develop an algorithm and its corresponding Python script for a game where the program asks the user's name. After a name is given the program generates a random integer between 1 and 20. The game should be in the following way (suppose that the generated number is 16):

   ```
   Hello, what is your name?
   John
   Very well John, Gues the number between 1 and 20.
   6
   Your number is too low.
   Give me another number
   12
   Your number is too low.
   Give me another number
   16
   Congratulations, you have guessed the number in 3 tries.
   ```

3. Write a script that reads a phrase. The output must be the count for each letter in the phrase.

4. In the previous exercise order the result in alphabetical order.

5. Generate a string consisting of 10 braces (either [or]). Devise a Python program that checks how many pairs are correct, for example, the following combinations can be considered:

[]	correct
]]	not correct
[[]]	correct
[] []	correct
[[] []]	correct
[[]]]]	not correct

6. The value of a word is the sum of the values assigned to a letter. If the value for each letter is as follows: a = 1, b = 2, c = 3, etc., design an algorithm and the corresponding Python script to compute the value of a received word. Use a dictionary to assign the value for each letter as:

   values = { "a": 1, "b": 2, ..., "z" :26 }

7. What is the result in the following loop applied to the Galileo's phrase:

   text = "The alphabet of the Universe is the Mathematics."

   ```python
   for k in text.split():
       print(k)
   ```

8. Repeat the previous exercise with:

   ```python
   for k in text.split('t'):
       print(k)
   ```

9. Add the strings "23" and "58" and convert the resulting string to a floating point number.

10. Write a Python script to read four words, compute the string length and display the words and the length indicated by asterisks. For example:

memory	******
mouse	*****
computer	********
dots	****

11. Convert the string "Computer" to a list and to a tuple.

12. Implement an algorithm and the corresponding script to implement the Caesarean code. The rule for this code is to change each letter in a text by another letter either before or after the given letter in the alphabet by a given number of positions. For example, if two positions after the letter are used, then the letter a is changed to c, the f is changed to h, and so on. The text must be read with an input instruction.

13. Write a script to code a word to its equivalent in the International Civil Aviation Organization (ICAO) phonetic code. The ICAO code must be stored in a dictionary in the following way:

code = {'A' : 'Alfa', 'B' : 'Bravo', 'C' : 'Charlie', 'D' : 'Delta', 'E' : 'Echo', 'F' : 'Foxtrot', 'G' : 'Golf', 'H' : 'Hotel', 'I' : 'India', 'J' : 'Juliett', 'K' : 'Kilo', 'L' : 'Lima', 'M' : 'Mike', 'N' : 'November', 'O' : 'Oscar', 'P' : 'Papa', 'Q' : 'Quebec', 'R' : 'Romeo', 'S' : 'Sierra', 'T' : 'Tango', 'U' : 'Uniform', 'V' : 'Victor', 'W' : 'Whisky', 'X' : 'Xray', 'Y' : 'Yankee', 'Z' : 'Zulu' }

14. One application of sets is to get rid of duplicates. Use a for loop to generate a random set of integer numbers between 1 and 10. Run the loop for 100 loops.

Chapter 4

Arrays

4.1 Introduction

An array is a collection of objects. These objects usually are of the same type, but this is not always the case. Arrays are used in a great deal of applications, for example, to store and handle bank accounts, census information, inventories, sports statistics, and payrolls, to name a few.

The most common arrays are vectors and matrices. In Python, arrays are represented by lists. In this way, a list represents a vector, a list of lists represents a two-dimensional array, and so on. In this chapter, the module numpy is frequently used because it has a great deal of functions that are adequate to handle arrays. *Pandas* is a package that handles arrays with functions similar to the ones used with Numpy but that has found many applications in data science and machine learning.

4.2 Introduction to Arrays

An array is a finite collection of data. For example, it may be a collection of integers, real numbers, alphanumeric variables, or even a combination of them. Thus, an array can be defined as:

An array is a finite collection of data.

- It is finite because the size is known.

- The elements type may be of the same type or of a different type. For example, some elements may be integer numbers, and some other elements may be floating-point numbers, some other elements may be strings, and some elements may be lists.

DOI: 10.1201/9781003222118-4

An array must have a name and a size; that is, the number of elements. Although the number of elements may be known beforehand, this is not always the case, as it will be seen in the chapter.

4.3   Vectors

The simplest array is a vector represented by a list. Thus, a list may be thought of as a vector. The following lists are vectors:

```
speeds = [ 55, 60, 75, 80]
trees = ['willow', 'aspen', 'elm', 'spruce', 'oak']
geometric_figures = ['triangle', 3, 'square', 4, 'rectangle', 5]
```

In the vector `speeds` the elements are integer numbers, in the vector `trees` the elements are strings, but in the vector `geometric_figures` the elements are strings and integer numbers. Thus, the elements in a vector or in a list can be of heterogeneous type.

An array can also be created with `numpy`. Importing `numpy` as `np`, the vector `speeds` is generated with:

```
import numpy as np
speeds2 = np.array([ 55, 60, 75, 80])
```

Looking for the type of both `speeds` and `speeds2`, it is seen that:

```
>>> type(speeds)
<class 'list'>

>>> type(speeds2)
<class 'numpy.ndarray'>
```

`ndarray` means that the array is multidimensional. In an array, the type of the elements must be the same. Numbers and strings cannot be mixed. If some elements are numbers and other elements are strings, `numpy` converts every element to a string, as in the vector `a90` given by:

```
>>> a90 = np.array([1, 2, 'string'])
>>> print(a90)
array(['1', '2', 'string'], dtype = '<U11')
```

It can be seen that `numpy` has converted the numbers 1 and 2 to strings. In the `dtype` the sign `<U11` means that the maximum size of each string is 11. To return an array to a list use the instruction:

TABLE 4.1:   Grades.

Index	Id_number	Grade
0	301834	9.3
1	301846	9.9
2	301847	8.6
3	301889	10.0
4	301996	9.6
5	301998	8.9
6	302020	10.0
7	302028	9.1
8	302067	10.0
9	302135	10.0

```
name_of_the_array.tolist()
```

For example, for the array `speeds2 = np.array([55, 60, 75, 80])`, it becomes a list with:

```
>>> speeds3 = speeds2.tolist()
>>> type(speeds3)
< class 'list'>
```

It can be seen that `speeds3` is a list.

Example 4.1 Array for students' grades
Grades for students are stored in vector `grades`. There is another vector with id numbers `id_numbers`. Data for these two vectors is shown in Table 4.1. The column to the right contains the grades, the column at the center contains the id numbers, and the column to the left contains the index for each student. Each vector has 9 elements. As with lists, the index indicates the position of an element in the vector. The type of each vector is integers for the index and `id_numbers` vector and floating-point for the `grades` vector. The vectors `id_numbers` and `grades` are then

```
grades = [9.9, 8.6, 10.0, 9.6, 8.9, 10.0, 9.1, 10.0]
id_numbers = [301834, 301846, 301847, 301889, 301996, 301998,
              302020, 302028, 302067]
```

These vectors can also be defined with `numpy` as:

```
import numpy as np
grades = np.array([9.9, 8.6, 10.0, 9.6, 8.9, 10.0, 9.1, 10.0])
```

```
id_numbers = np.array([301834, 301846, 301847, 301889, 301996,
        301998, 302020, 302028, 302067]
```

CAUTION

Vectors defined with **numpy** are not lists and thus, the functions for lists cannot be applied to **numpy** arrays.

It can readily be seen that the student with id 302020 has a grade of 10.0 whereas the student with id number 301834 has a grade 9.9. In the first student, the index is 6 and in the second one the index is 0.

Each element of the vector is called an element or a component. Each component of a vector is referenced by its index. For example, **id_number[1]** indicates the student with the No. 301846 and **grades[2]** indicate 8.6.

A **for** loop can be used to print out the id numbers and the grades. The function **range** can be substituted by the equivalent function in **numpy** which is **arange** as:

```
for student in np.arange(9):
        print(id_numbers[student], "\t", grades[student])
```

and the result is:

301834	9.3
301846	9.9
301847	8.6
301889	10.0
301996	9.6
301998	8.9
302020	10.0
302028	9.1
302067	10.0

In this **for** loop the index is **student** and it takes the values from 0 to 8 for a total of 9 id numbers and 9 grades.

Example 4.2 Average of the elements of a vector

It is desired to calculate the average of the grades. This is obtained by adding the grades and dividing the result by the number of grades which in this example is 9. The sum has to be initialized to 0, then a **while** loop is used to do this:

```
grades = [9.9, 8.6, 10.0, 9.6, 8.9, 10.0, 9.1, 10.0]
sum = 0
```

```
i = 0 # This is the index.
while i < 9:
      sum = sum + grades[i]
      i = i + 1 # The index is incremented by 1.
```

The sum is now divided by the number of grades. The value of the index i is the number of grades. Thus:

```
average = sum/i
# The average is displayed:
print('The grade average is: %0.2f' %average)
```

The complete script is:

```
"' This is script Example_4_2.py
This program finds the average of the grades.'"

# The grades are:
grades = [9.9, 8.6, 10.0, 9.6, 8.9, 10.0, 9.1, 10.0]

# Initialize the sum
sum = 0
i = 0 # This is the index.
while i < len(grades):
      sum = sum + grades[i]
      i = i + 1 # The index is incremented by 1.

average = sum/i
# The average is displayed:
print('The grade average is: %0.2f' %average)
```

The result for the grade average for this vector of grades is:

```
The grade average is: 9.5125
```

There are two functions in numpy that compute the average, they are np.average and np.mean. For the vector in this example:

```
>>> import numpy as np
>>> grades = [9.9, 8.6, 10.0, 9.6, 8.9, 10.0, 9.1, 10.0]
>>> print("mean", np.mean(grades))
>>> print("average", np.average(grades))
```

The result is:

```
mean 9.5125
average 9.5125
```

4.3.1 Access to Vectors

As seen from the previous examples, vector elements can be accessed by the index and not by the value. Thus, if a grade is to be changed, the user has to do it using the index. For example, the index is used if the grade for a student with id number 301998 has to change to 9.1. For this student, the index is 5 and then:

```
grades[5] = 9.1
```

When using the index in a vector, in a list or in an array (as well in a string, tuple, or dictionary) the index must be smaller than the length of the vector. In the case that an index is out of these bounds, Python emits an error message. For example, if it is desired to print out **grades[11]**, then:

```
print(grades[11])
```

The error message is:

```
Traceback (most recent call last):
   File "<pyshell#5>", line 1, in <module>
   print(grades[11])
IndexError: list index out of range
```

4.3.2 Vectors by Comprehension

Defining an array by comprehension means that it can be defined in a single expression. Vectors and lists can be defined by comprehension as:

```
[f(x) for x in range(A)]
```

This function generates a vector $f(x)$ for x-values $x = 0, 1, 2, \ldots,$ A-1. The vector is:

$$[f(0), f(1), \ldots, f(\text{A-1})]$$

For example, a vector of squared elements is:

```
random_vector = [x**2 for x in range(5)]
```

with the result:

```
[0, 1, 4, 9, 16]
```

4.3.3   The Instruction append

New elements can be added to a vector with the instruction **append**. This instruction was used in Chapter 2 for lists. If it is desired to add an element x = -27 to vector a = [2, 5, 6, -3], this can be done with

```
>>> a = [2, 5, 6, -3]
>>> a.append(-27)
>>> print("a = ", a)
```

and the result is

```
a = [2, 5, 6, -3, -27]
```

In the case of **numpy** arrays, the use of the instruction **append** is different and it is used as:

```
>>> import numpy as np
>>> a = np.array([-3, 5, 8, 27])
>>> a1 = np.append(a, 99)
>>> print(a1)
```

The resulting vector a1 is:

```
[-3  5  8  27  99]
```

Note that the elements are not separated by commas.

4.4   Examples with Vectors in Python

In this section, some examples with vectors are presented. Since vectors are lists, the examples also apply to lists.

Example 4.3 Vector with random numbers

The module or library **random** contains functions to generate random numbers. The function **random** in this library generates random numbers between 0 and 1. The function has to be imported before using it. As an example:

```
>>> import random
>>> random.random( )
```

The result is:

```
0.5338963331348068
```

which is a number between 0 and 1. To generate integer random numbers, the function **randint(min, max)** generates integer random numbers between **min** and **max**. For example, to generate integer random numbers between 23 and 32:

```
a = random.randint(23, 32)
```

produces an integer random number:

```
25
```

In both cases, **the random numbers generated can differ depending upon the computer and the time.** To generate a vector with ten random numbers first it is needed to import the library **hyphen** and initialize the vector as:

```
import random
x = [ ]
```

Then, the number of random numbers N have to be read:

```
N = int(input("Enter the number of random numbers: "))
```

A **for** loop is used to generate the N random numbers and the function **append** is used to append them to the vector x:

```
for index in range(N):
    x.append(random.random())
```

Finally, the vector x is printed out:

```
print("The vector x is: ")
print (x)
```

The complete script is:

```
‘‘‘ This is script Example_4_3.py
This program generates a vector of random numbers.

The library random is imported, the vector is initialized:’’’
import random
x = [ ]

# The number of random numbers N is read:
N = int(input(“Enter the number of random numbers: ”))

# A for loop generates the vector x:
for index in range(N):
    x.append(random.random())

# The vector x is printed out:
print(“The vector x is:”)
print(x)
```

A run for N = 7 produces:

```
Enter the number of random numbers: 7
The vector x is:
[0.5915808827906403, 0.6249345836764961, 0.9723693863413662,
0.7841691965848496, 0.7599104145837037, 0.14857081490159296,
0.2035987794727152]
```

The vector is printed horizontally. If the vector is to be printed vertically, the
print instruction must be included in the for loop. The program is modified
as:

```
‘‘‘ This is script Example_4_3b.py
This program generates a vector of random numbers.
The library random is imported, the vector is initialized:’’’

import random
x = [ ]
# The number of random numbers N is read:
N = int(input(“Enter the number of random numbers: ”))

# Print the heading:
print(“The vector x is:”)

# Example 4.3b continues next page.
```

```
# Example 4.3b continued from previous page.

# A for loop generates the vector x:
for index in range(N):
     x.append(random.random())
     print(x[index])

# After the for loop the vector is stored in the vector x.
```

A run for N = 7 gives:

```
Enter the number of random numbers: 7
The vector x is:
0.05424600766075338
0.15136275059034443
0.35408879989961306
0.48668016182067797
0.6638698893544993
0.7752464332133799
0.16854516883756054
```

Example 4.4 Vector with N integer random numbers

To generate a vector with N integer random numbers, as before, first, it is needed to import the library **random** and initialize the vector. It is necessary to read the maximum and minimum limits of the random integers. The function **randint()** generates the random integers. In this example, the vector size N, the maximum, and the minimum are read and then the vector x is initialized as:

Read N, max, and min:

```
N = int(input("Enter the size of the vector: "))
max = int(input("Enter the maximum value: "))
min = int(input("Enter the minimum value: "))
```

Initialize the vector x:

```
x = [None]*N
```

Print out the heading:

```
print("The integer random numbers are: ")
```

A `for` loop generates the integers:

```
for k in range(N):
    x [k] = random.randint(min, max)
    print(x[k])
```

The complete script is:

```
"' This is script Example_4_4.py
This program generates a vector of integer random numbers.

The library random is imported:'"
import random

# Read N, max, and min:
N = int(input("Enter the size of the vector: "))
max = int(input("Enter the maximum value: "))
min = int(input("Enter the minimum value: "))

# Initialize the vector x:
x = [None]*N

# Print out the heading:
print("The integer random numbers are: ")

# A for loop generates the integers:
for k in range(N):
    x[k] = random.randint(min, max)
    print(x[k])
```

A run with $N = 6$, $min = 30$, and $max = 50$ produces the result:

```
33
46
39
40
36
47
```

The random numbers can change depending on the computer used and the time of the execution.

Example 4.5 Vector with N floating random numbers
The previous example generated integer random numbers. To generate a vector with N floating random numbers, it is necessary to take the module of the

random number generated with **random**, and it is known that it is less than unity by a factor given by the maximum and minimum of the range as:

```
x[k] = min + max*random.random()%(max - min)
```

This results in a floating random number in the range between **min** and **max**. This is the only change that has to be done in Example 4.3. The final program is:

```
"' This is script Example_4_5.py
This program generates a vector of random numbers in a range.

The library random is imported:'"
import random

# Read N, max, and min:
N = int(input("Enter the size of the vector: "))
max = int(input("Enter the maximum value: "))
min = int(input("Enter the minimum value: "))

# Initialize the vector x:
x = [None]*N

# Print out the heading:
print("The random numbers are: ")

# A for loop generates the random numbers:
for k in range(N):
    x[k] = min + max*random.random()%(max - min)
    print(x[k])
```

A run is:

```
Enter the size of the vector: 6
Enter the maximum value: 31
Enter the minimum value: 21
The random numbers are:
25.968856257707415
28.703127598617424
25.595007189273403
27.80705467890867
30.116983247056243
24.903925442918176
```

Example 4.6 Average, variance, and standard deviation

In this example, the average, variance, and standard deviation of a vector are computed. The average, also known as mean value, was computed in Example

4.2 and is repeated here. The equations that define these three functions for a vector X[i] are for the average or mean value:

$$\overline{X} = \frac{1}{n} \sum_{i=0}^{n-1} X[i]$$

For the variance, also known as dispersion, it is a measure that the dispersion of the data points around its center of gravity given by the mean value. A large variance is an indication that many data points are distant from the mean value. On the other hand, a small value for the variance is an indication that the data points are clustered around the mean value. The variance is represented by σ^2 with:

$$\sigma^2 = \text{Variance} = \frac{1}{n} \sum_{i=0}^{n-1} \left(X[i] - \overline{X}\right)^2$$

Finally, the standard deviation σ is defined by the square root of the variance. The variance and the standard deviation have the same meaning, but the standard deviation has the same units as the elements of the vector. The standard deviation σ is given by:

$$\sigma = \text{standard deviation} = \sqrt{\text{Variance}}$$

For a given vector X of length N the average, variance and standard deviation can be computed. For the average:

```python
# Initialize the sum
sum = 0
i = 0 # This is the index.
while i < len(grades):
    sum = sum + grades[i]
    i = i + 1 # The index is incremented by 1.

average = sum/i
```

The variance requires the mean value. The following script computes the variance. First, the sum for the variance is initialized as:

```python
variance = 0.0
```

The sum is done with a **for** loop and the variance is obtained by dividing the sum by N:

```python
for i in range(N):
    sum = sum + (X[i] - average)**2

variance = sum/N
```

The standard deviation is the square root of the variance, then:

```
sigma = sqrt(variance)
```

Finally, the average, variance, and standard deviation are printed out:

```
print('The average or mean value is: %0.2f' %average)
print('The variance is: %0.2f' %variance)
print('The standard deviation is: %0.2f' %sigma)
```

The complete script is:

```
''' This is script Example_4_6.py
This program computes the average, variance, and standard
deviation.

The function sqrt from the library math is imported:'''
from math import sqrt

# Read the length of the vector X:
N = int(input("Enter the length of the vector X: "))
X = [ ]

# Read the numbers for vector X:
for i in range(N):
    element = float(input("Enter a value for vector X: "))
    X.append(element)
# Compute the average:
# Initialize the sum
sum = 0
i = 0 # This is the index.
while i < len(X):
    sum = sum + X[i]
    i = i + 1 # The index is incremented by 1.

average = sum/i

# Initialize the sum for the variance:
variance = 0.0

# Example 4.6 continues at next page.
```

```
# Example 4.6 continued from previous page.

# Start the for loop for the sum in the variance:
for i in range(N):
    sum = sum + (X[i] - average)**2

variance = sum/N # variance

# Standard deviation:
sigma = sqrt(variance)

# Display the average, variance, and standard deviation:
print('The average or mean value is: %0.2f' %average)
print('The variance is: %0.2f' %variance)
print('The standard deviation is: %0.2f' %sigma)
```

As an example, for vector X = [1, 3, 5, 7, 9, 11, 13, 17, 19, 21], the length of the vector is N = 10, the average, variance, and standard deviation are:

```
The average or mean value is: 10.60
The variance is: 52.84
The standard deviation is: 7.27
```

4.4.1 Vector Sorting

The elements of a vector can be ordered either in ascending or descending order depending upon the user needs. This process is known as sorting. There are several methods available for sorting vectors such as selection, bubble, merge, and insertion, to name a few methods. Python includes a method which is a combination of some of these methods and is very efficient in terms of memory usage and speed. This method is known as the **timsort** algorithm and is a combination of the insertion and merge algorithms. It was created by Tim Peters in 2002. A detailed description of the algorithm is beyond the scope of this book but the interested reader can see reference 1 for details. The algorithm **timsort** can be called with the function **sort**.

Example 4.7 Sorting of a random vector

It is desired to generate a vector with N random integer numbers between min and max. Then the *timsort* algorithm is used for ordering it. The first step is to import the random module and to read the number of elements in the vector, and the minimum and maximum values. The vector is initialized also. This is done with:

```
from random import randint
N = int(input('Enter the number of elements: '))
min = int(input('Enter the minimum value for the elements: '))
max = int(input('Enter the maximum value for the elements: '))
vector = [None]*N
```

In a `for` loop generate the vector elements as:

```
for i in range(N):
    vector[i] = randint(min, max)
```

The original vector is displayed:

```
print("Original vector")
print(vector)
```

The sorting algorithm now is used:

```
vector.sort() # Vector is sorted.
sorted_vector = vector
```

The sorted vector is displayed:

```
print("Sorted vector")
print(sorted_vector)
```

The final program is:

```
"' This is script Example_4_7.py
This program sorts a random vector."'

# Import the function randint from the module random:
from random import randint

# Read the number of elements N and the minimum and maximum:
N = int(input('Enter the number of elements: '))
min = int(input('Enter the minimum value for the elements: '))
max = int(input('Enter the maximum value for the elements: '))
vector = [None]*N

# Loop to generate the vector elements:
for i in range(N):
    vector[i] = randint(min, max)

# Script continues at next page.
```

```
# Script continued from previous page.

# The original vector is displayed:
print("Original vector")
print(vector)

# The sorting algorithm is now used:
vector.sort() # Sorted vector
sorted_vector = vector

# The sorted vector is displayed:
print("Sorted vector")
print(sorted_vector)
```

A run is:

```
Enter the number of elements: 7
Enter the minimum value for the elements: -5
Enter the maximum value for the elements: 9
Original vector
[5, 3, 6, 3, 1, -3, -1]
Sorted vector
[-3, -1, 1, 3, 3, 5, 6]
```

4.5   Matrices

Matrices are data structures that frequently appear in mathematics. They can represent system models such as the coefficients of a system of simultaneous linear equations, the pixels in an image, and a neural network, to name a few applications. An example of the use of matrices is in the handling of sales in a department store. Here, a vector may represent the sales in a month, but for the whole year, it is more convenient to use a matrix instead of 12 vectors. Another example is when forces act on a structure.

A matrix is a multidimensional array that can be used to store data. A two-dimensional array is composed of rows and columns as the following array:

$$
A = \begin{bmatrix}
a_{0,0} & a_{0,1} & \cdots & a_{0,n-1} \\
a_{1,0} & a_{1,1} & \cdots & a_{1,n-1} \\
\vdots & & & \\
a_{m-1,0} & a_{m-1,1} & \cdots & a_{m-1,n-1}
\end{bmatrix}
$$

A parameter of interest in a matrix is the dimension which is related to the size of the array. The array A above has m rows and n columns, arrays such as this are called two-dimensional and the dimension is $m \times n$. Two-dimensional matrices are very common such as the case of a gray scale image where each picture element, called a pixel, is located with two coordinates.

Each element of the matrix has a position indicated by the number of the row and the number of the column. Thus, the element of the array A which is located in the p-th row and q-th column is referenced as:

$$a_{p,q}$$

or simply

$$a_{pq}$$

Example 4.8 Examples of two-dimensional arrays

Consider the matrices A, B, y C given by:

$$A = \begin{bmatrix} \text{``}a\text{''} & \text{``}b\text{''} \\ \text{``}c\text{''} & \text{``}d\text{''} \\ \text{``}e\text{''} & \text{``}f\text{''} \end{bmatrix}, \quad B = \begin{bmatrix} 77.3 & 3.1416 \\ 10.01 & -201.3 \\ 66.6 & 27.021 \end{bmatrix}, \quad C = \begin{bmatrix} 1 & 2 & 3 \\ 9 & 18 & -5 \\ 16 & 4 & 7 \end{bmatrix}$$

Matrix A is composed of alphanumeric characters and its size is 3×2. Array B has three rows and two columns and thus it is a 3×2 floating-point matrix. Finally, matrix C is a square matrix of dimension 3×3.

4.6 Arrays in Python

Arrays in Python are described as a **list** of **lists**. A matrix such as:

$$A = \begin{bmatrix} 1 & -2.3 & 0 \\ 9.6 & -1.5 & 4 \\ 5.7 & -6.3 & 3.1 \end{bmatrix}$$

is represented in Python as:

```
A = [ [ 1, -2.3, 0], [ 9.6, -1.5, 4 ], [ 5.7, -6.3, 3.1 ] ]
```

The list is A and has three elements. The first list [1, -2.3, 0] is the first element (element 0) of list A. The following list [9.6, -1.5, 4] is the element 1 of list A. Finally, the list [5.7, -6.3, 3.1] is the element 2 of list A. Each one of these lists has three elements.

The array `A` can be printed with `print(A)` which produces the list as shown above. If it is desired to print out the matrix as a table, two nested `for` loops can be used. The first `for` loop prints a row and the second one prints the following row:

```
for i in range(3): # i is for rows.
    print( ) # Used for the next row.

    for j in range(3): # j is for columns.
        print(A[i][j], end = '\t ')
```

The result is:

1	-2.3	0
9.6	-1.5	4
5.7	-6.3	3.1

An escape sequence for a tab given by '\t' has been added in the last print for a better presentation.

4.6.1   Array Generation by Indexing

To generate an array sometimes it is best to generate a blank array and then fill it up with the required data. A `for` loop can be used to generate the array with the desired size and then fill it up with the required data. To do this, a `for` loop is used and the `index` changes value; thus, the name indexation. This can be done as:

1. Generate an empty list. This is done with:

   ```
   A = [ ]
   ```

2. With a `for` loop generate the list corresponding to the rows. The length of these lists is the number of columns `m`:

   ```
   for i in range(m):
           A.append([ ]) # A row is generated.
           for j in range(n):
                   A[i].append(None) # Generates each one of the elements.
   ```

The matrix can be displayed with a pair of nested `for` loops as:

```
for k in range(m): # Index for the rows.
    print( )   # Jump to another row.
```

```
for j in range(n): # Index for the columns.
    print( A[k][j] ," ," \t )
```

Example 4.9 Generation of a matrix

It is desired to have a $m \times n$ matrix. This can be done as:

Read the number of rows and columns:
```
m = int(input("Enter the number of rows: ")
n = int(input("Enter the number of columns: ")
```

Initialize the matrix:

```
A = [ ]
```

Generate the rows and columns:

```
for i in range(m):
    A.append([ ])
    for j in range(n):
        A[i].append([None])
```

Display the matrix:

```
for k in range(m):
    print( ) # New row.
    for j in range(n):
        print(A[k][j], "\t")
```

The final script is:

```
"' This is script Example_4_9.py
This program generates a mxn matrix."'

# Read the number of rows and columns:
m = int(input("Enter the number of rows: ")
n = int(input("Enter the number of columns:  ")

# Initialize the matrix:
A = [ ]

# Example 4.9 continues at next page.
```

```
# Example 4.9 continued from previous page.

# Generate the rows and columns:
for i in range(m):
    A.append([ ])
    for j in range():
        A[i].append([ ])

# Display the matrix:
for k in range(m):
    print ( )# For new row.
    for j in range(n):
        print(A[k][j] , end = "\t")
```

The matrix obtained for m = 4 and n = 4 is:

```
[ ] [ ] [ ] [ ]
[ ] [ ] [ ] [ ]
[ ] [ ] [ ] [ ]
[ ] [ ] [ ] [ ]
```

4.6.2   Array Generation by Comprehension

Comprehension can be used to generate two-dimensional arrays when the elements of the array are defined by a function. Basically, this is equivalent to two nested **for** loops. The first loop generates the rows and the nested loop generates the columns. The format is:

```
A = [[f(i) for i in range(No. columns)] for j in range(No. rows)]
```

In this definition, the rows are identical but the goal is to generate the array with the desired size. For example, a 6×5 array with 0's can be generated with:

```
A = [[0 for i in range(5)] for j in range(6)]
```

A random function can also be used. For example:

```
from random import *
A = [[randint(-10, 10) for i in range(5)] for j in range(6)]
```

and the result is:

```
[[-7, 8, 5, 6, 9], [-10, 6, -6, -2, -3], [-7, -6, 10, -5, 1],
[1, -4, 6, 10, 3], [-10, -6, 4, 3, 4], [0, 8, 8, 9, 1]]
```

This matrix is randomly generated and the results change for each run.

Example 4.10 Use of a list iterator
A list iterator refers simply to row handling in a matrix. For example, for the matrix:

$$
\texttt{matrix} = \begin{bmatrix} -1 & 0 & -10 \\ 3 & -4 & 0.5 \\ 6 & -23 & 8 \\ 7 & 3 & 9 \end{bmatrix}
$$

can be written in Python as:

```
matrix = [[-1, 0, -10], [3, 4, 0.5], [6, -23, 8], [7, 3, 9]]
```

To display it, a list iterator can be used:

```
for row in matrix:
    print(row)
```

The result is:

```
[ -1,     0,     -10 ]
[ 3,      4,      0.5 ]
[ 6,      -23,      2 ]
[ 7,      3,        9 ]
```

Example 4.11 Another way of list iterators
Another way of list iterators is the following:

```
for row in matriz: # Selects the rowss.
        print( ) # For a new row.
        for c in row:# Selects the columns.
            print(c, end = ', ')# Comma and space to separate the
columns.
```

The result is:

```
-1,     0,     -10,
3,      4,      0.5,
6,      -23,    8,
7,      3,      9,
```

TABLE 4.2:   Operations for numpy arrays

Operator	Function
`array.tolist()`	converts an array to list
`array(list)`	converts a list to array
`dot(v1, v2)`	dot product of vectors v1, v2
`vdot(v1, v2)`	dot product of vectors v1, v2
`inner(v1, v2)`	dot product of vectors v1, v2
`linalg.eig(A)`	eigenvectors of matrix A
`linalg.matrix_power(A,n)`	A^n power
`linalg.norm(v, N)`	norm N of a vector
`linalg.det(A)`	determinant of A
`linalg.trace(A)`	trace of A
`linalg.inv(A)`	inverse of A
`linalg.solve(A, b)`	solve system of linear equations
`linalg.transpose(A)`	transpose of A

It can be noted that there are no brackets.

4.7   Matrix Operations using Linear Algebra with numpy

The library numpy includes a set of functions for linear algebra in the module `linalg` which are ideal to compute functions on matrices. Recall that a matrix is an array for numpy. Vectors are also arrays in numpy and thus, many operations are available for vectors. A list is shown in Table 4.2. A set of examples follow.

4.7.1   Sum, Difference, and Multiplication of Matrices

For arrays the following operations exist:

- Sum and difference of matrices of the same dimension.

- The product of a scalar by a matrix.

- The product of a matrix times another matrix.

4.7.2   Sum of Matrices

The sum and difference of matrices can only be done when both matrices or arrays have the same dimensions; that is, the same size. At least one of them has to be an array. If both arrays are lists then a concatenation is performed.

```
>>> import numpy as np
>>> a = np.array([[1, 2], [4, 3]])
```

```
>>> b = [[5, 6], [7, 8]]
>>> c = a + b
>>> print(c)
```

The result is:

```
[[ 6 8]
 [11 11]]
```

4.7.3 Product of a Matrix by a Matrix

The product of a matrix A of dimension $n \times m$ times another matrix B of dimension $j \times k$ requires that the dimensions of the matrices be such that $m = j$. This means that the number of columns of matrix A is equal to the number of rows of matrix B. **The result is a matrix of dimension** $n \times k$. For the matrices:

$$A = \begin{bmatrix} a_{0,0} & a_{0,1} & \cdots & a_{0,m-1} \\ a_{1,0} & a_{1,1} & \cdots & a_{1,m-1} \\ \vdots & & & \\ a_{n-1,0} & a_{n-1,1} & \cdots & a_{n-1,m-1} \end{bmatrix}$$

$$B = \begin{bmatrix} b_{0,0} & b_{0,1} & \cdots & b_{0,k-1} \\ b_{1,0} & b_{1,1} & \cdots & b_{1,k-1} \\ \vdots & & & \\ b_{m-1,0} & b_{m-1,1} & \cdots & b_{m-1,k-1} \end{bmatrix}$$

The elements of matrix $C = A*B = AB$ are given by:

$$c_{ij} = \sum_{l=0}^{m-1} a_{il}b_{lj} = a_{i0}b_{0j} + a_{i1}b_{1j} + \cdots + a_{i,m-1}b_{m-1,j}$$

where the indices i, j vary from

$$i = 0, 1, 2, \ldots, n-1 \qquad j = 0, 1, 2, \ldots, k-1 \qquad (4.1)$$

This is equivalent to say that the element c_{ij} is obtained by multiplying the elements of the row i of matrix A by the corresponding elements of column j of matrix B and adding the products. The only requirement is to multiply two matrices is that **the number of columns of the first matrix is equal to the number of rows of the second matrix.**

Example 4.12 Matrix multiplicacion

For matrices A and B given by:

$$A = \begin{bmatrix} 1 & 2 & 3 \\ 4 & 5 & 6 \end{bmatrix}, \quad B = \begin{bmatrix} -7 & -8 \\ 9 & 10 \\ 4 & -11 \end{bmatrix}$$

The order of A is 2×3 and the order of B is 3×2. Then, the product AB can be obtained to be a matrix of order 2×2. Also, the product BA can be obtained as a matrix of order 3×3. These products can be done with:

$$AB = \begin{bmatrix} 1 & 2 & 3 \\ 4 & 5 & 6 \end{bmatrix} \begin{bmatrix} -7 & -8 \\ 9 & 10 \\ 4 & -11 \end{bmatrix}$$

$$= \begin{bmatrix} 1*(-7)+2*9+3*4 & 1*(-8)+2*10+3*(-11) \\ 4*(-7)+5*9+6*4 & 4*(-8)+5*10+6*(-11) \end{bmatrix}$$

$$= \begin{bmatrix} 23 & -21 \\ 41 & -48 \end{bmatrix}$$

which is a matrix of order 2×2. Also:

$$BA = \begin{bmatrix} -7 & -8 \\ 9 & 10 \\ 4 & -11 \end{bmatrix} \begin{bmatrix} 1 & 2 & 3 \\ 4 & 5 & 6 \end{bmatrix}$$

$$= \begin{bmatrix} (-7)*1+(-8)*4 & (-7)*2+(-8)*5 & (-7)*3+(-8)*6 \\ 9(1)+10*4 & 9*2+10*5 & 9*3+10*6 \\ 4*1+(-11)*4 & 4*2+(-11)*5 & 4*3+(-11)*6 \end{bmatrix}$$

$$= \begin{bmatrix} -39 & -54 & -69 \\ 49 & 58 & 87 \\ -40 & -47 & -54 \end{bmatrix}$$

which is a matrix of order 3×3.

4.7.4 Product of Matrices in Python

There are two techniques to multiply matrices in Python and in both cases, numpy functions are used. The first of these techniques uses the numpy function np.dot. The other one uses the classical * to multiply the matrices but in this case, the arrays have to be converted to numpy matrices using np.matrix(). To convert back from np.matrix to list the instruction np.matrix.tolist(name_of_the_matrix).

4.7.4.1 Matrix Multiplication Using np.dot

Using numpy it is very easy to do matrix multiplication if the matrices are defined as arrays. In that case, the product of matrices can be obtained with the dot function available in numpy. The product of an $m \times n$ matrix A and an $n \times p$ matrix B produces a matrix C of dimension $m \times p$ as in the following matrices:

```
>>> import numpy as np
>>> A = np.array([[2, 3], [5., 2.], [0, -7]]) # Dimension 3x2.
>>> B = np.array([[7, 8.0], [-6., -9.]]) # Dimension 2x2
>>> np.dot(A, B)
```

and the result is a 3×2 array given by:

```
array([[ -4., -11.],
       [ 23., 22.],
       [ 42., 63.]])
```

4.7.4.2 Matrix Multiplication Using np.matrix

Another way to implement matrix multiplication is converting the list of lists or numpy array to a numpy matrix. The format is np.matrix(). In this case, the product is obtained as A*B. The previous arrays are now multiplied as:

```
>>> import numpy as np
>>> A = np.matrix([[2, 3], [5., 2.], [0, -7]])
>>> B = np.matrix([[7, 8.0], [-6., -9.]])
>>> A*B
```

The result is now a 3×2 matrix which is the same result as before:

```
matrix([[ -4., -11.],
        [ 23., 22.],
        [ 42., 63.]])
```

Example 4.13 Matrix multiplication using numpy

This example generates two matrices A and B with random integers and obtains the product C $=$ A*B. The data needed are the dimensions of the matrices. The following file reads the dimensions of three matrices and then generates them with random integers. The multiplication of the matrices is implemented as arrays with **np.dot** and when the matrices are declared with **np.matrix** the multiplication is implemented with the asterisk *.

```
'' This is script Example_4_13.py
This program multiplies two matrices a and b. The result is
matrix c. The dimension of the matrices is input data. The
multiplication is done as arrays using np.dot and as numpy
arrays.'"

import numpy as np
from random import *

m = int(input("Enter the number of rows of matrix a: "))
n = int(input("Enter the number of columns of matrix a: "))
p = int(input("Enter the number of columns of matrix b: "))

# Initialize the arrays
a = np.array([[0 for i in range(n)] for j in range(m)])
b = np.array([[0 for i in range(p)] for j in range(n)])
c = np.array([[0 for i in range(p)] for j in range(n)])

# Generate the matrices a and b with random integers:
for k in range(m):
    for j in range(n):
        a[k][j] = randint(0, 9)
for k in range(n):
    for j in range(p):
        b[k][j] = randint(0, 9)

# Display the arrays:
print('a = \n ', a)
print('b = \n ', b)

# Compute the product:
c = np.dot(a, b)
# Display the result of dot function.
print('Using np.dot, c = np.dot(a, b):\n c =', c)

# Script continues next page.
```

```
# Script continued from previous page.

# Display the result.
print('c = ', c)

a = np.matrix(a)
c = a*b
# Display the result of matrix multiplication with *.
print('Conversion of array a to matrix: c = a*b: \n c = ', c)
```

After running this program the result is:

```
Enter the number of rows of matrix a: 3
Enter the number of columns of matrix a: 4
Enter the number of columns of matrix b: 2

a =
[[9 5 9 1]
 [1 4 2 8]
 [1 6 0 4]]

b =
[[0 4]
 [1 5]
 [0 8]
 [9 5]]

Using np.dot, c = np.dot(a, b):

c =
[[ 14 138]
 [ 76 80]
 [ 42 54]]

Conversion of array a to matrix: c = a*b:

c =
[[ 14 138]
 [ 76 80]
 [ 42 54]]
```

The results will vary for each run because the matrices a, b, and c are created with a random generator.

4.8 Special Matrices

There are some matrices that deserve special treatment. These matrices are the identity, the transpose, the inverse, the triangular, the diagonal, among others.

4.8.1 The Identity Matrix

The identity matrix has 1's in the main diagonal and 0's everywhere else and it is a square matrix. It is represented by the letter I:

$$I = \begin{bmatrix} 1 & 0 & 0 & \cdots & 0 \\ 0 & 1 & 0 & \cdots & 0 \\ 0 & 0 & 1 & \cdots & 0 \\ \vdots & & & & \\ 0 & 0 & 0 & \cdots & 1 \end{bmatrix}$$

In Python, the identity matrix requires the library **numpy**. It is identified as **eye(n)** where **n** is the order and can be obtained as:

```
import numpy as np
A = np.eye(4) # Order 4 identity matrix.
print(A)
```

and the result is:

```
[[1. 0. 0. 0. 0.]
 [0. 1. 0. 0. 0.]
 [0. 0. 1. 0. 0.]
 [0. 0. 0. 1. 0.]
 [0. 0. 0. 0. 1.]]
```

4.8.2 The Transpose Matrix

The transpose matrix B is obtained by interchanging rows and columns of matrix A. If A has m rows and n columns, the transpose matrix has n rows and m columns. For matrix A given by:

$$A = \begin{bmatrix} a_{0,0} & a_{0,1} & \cdots & a_{0,n-1} \\ a_{1,0} & a_{1,1} & \cdots & a_{1,n-1} \\ \vdots & & & \\ a_{m-1,0} & a_{m-1,1} & \cdots & a_{m-1,n-1} \end{bmatrix}$$

the transpose matrix, denoted by $A^T = A'$, with n rows and m columns is given by:

$$A^T = A' = \begin{bmatrix} a_{0,0} & a_{1,0} & \cdots & a_{m-1,0} \\ a_{0,1} & a_{1,1} & \cdots & a_{m-1,1} \\ \vdots & & & \\ a_{0,n-1} & a_{1,n-1} & \cdots & a_{m-1,n-1} \end{bmatrix}$$

Another way to express the transpose of A is by using its elements a_{ij} as:

$$A = [a_{ij}] \qquad \text{Original matrix}$$

$$A^T = [a_{ji}] \qquad \text{Tranpose matrix}$$

As an example, given the matrix A:

$$A = \begin{bmatrix} -1 & 4 \\ 8 & 3 \\ 6 & 9 \\ 2 & -7 \end{bmatrix} \qquad \text{Original matrix}$$

The transpose is:

$$A^T = \begin{bmatrix} -1 & 8 & 6 & 2 \\ 4 & 3 & 9 & -7 \end{bmatrix} \qquad \text{Transpose matrix}$$

In Python, the transpose matrix can be obtained using the function `transpose()` from **numpy**. For example for matrix A, the transpose is obtained as:

```
At = np.transpose(A)
```

For **numpy** arrays, the transpose may also be obtained with **array.T**. Thus, for the array **A** above, the transpose can be found with **np.array(A).T**.

Example 4.14 Transpose of an array or list
Given the array :

```
A = [ [-1, 0, -10], [3, 4, 0.5], [6, -23, 8], [7, 3, 9] ]
```

The transpose matrix B can be obtained as:

```
>>> import numpy as np
>>> A = [ [-1, 0, -10], [3, 4, 0.5], [6, -23, 8], [7, 3, 9] ]
>>> At = np.transpose(A)
>>> print('At = \n', At)
```

```
>>>  At2 = np.array(A).T
>>>  print('At2 = \n', At2)
```

And the result is:

```
At =
[[ -1.  3.  6.  7.]
 [  0.  4.  -23 3.]
 [ -10.  0.5.  8.  9.]]

At2 =
[[ -1.  3.  6.  7.]
 [  0.  4.  -23 3.]
 [ -10.  0.5.  8.  9.]]
```

4.8.3   Transpose by Comprehension

The transpose of array A can be obtained by comprehension as:
`[[row[i] for row in A] for i in range(No_columns_of_A)]`

Example 4.15 Transpose of an array by comprehension
Consider the array:

$$A = \begin{bmatrix} -1 & 4 & 10 & 5 \\ 8 & 3 & -2 & -3 \\ 6 & 9 & 5 & -1 \\ 2 & -7 & 0 & 1 \end{bmatrix} \qquad \text{Original matrix}$$

In Python it is expressed as:

```
>>>  import numpy as np
>>>  A = np.array([[-1, 4, 10, 5], [8, 3, -2, -3], [6, 9, 5, -1],
[2, 7, 0, 1]])
```

For this array the transpose is

```
>>>  A_trans = [[row[i] for row in A] for i in range(len(A[0]))]
>>>  print(A_trans)
```

And the result is

```
[[-1, 8, 6, 2], [4, 3, 9, 7], [10, -2, 5, 0], [5, -3, -1, 1]]
```

This result is the matrix:

$$A\_trans = \begin{bmatrix} -1 & 8 & 6 & 2 \\ 4 & 3 & 9 & 7 \\ 10 & -2 & 5 & 0 \\ 5 & -3 & -1 & 1 \end{bmatrix}$$

4.9 Examples

In this section, two examples using matrices are presented. The first one is a throw of a dice and an array is used to store the throws. The second example solves a set of simultaneous linear differential equations using the eigenvalues and eigenvectors of the system matrix.

Example 4.16 Throw of dice

Three friends have a six-sided die. The die is thrown five times by each of them. The throws are stored in a 3×6 array where the last column is the total sum of points. For example,

Name	Throw 1	Throw 2	Throw 3	Throw 4	Throw 5	Total
John	6	3	4	4	3	20
Robert	4	5	6	4	5	24
Alfred	2	4	1	6	2	15

It is desired to write a Python program to implement the following actions:

1. Generate the 3×6 array.

2. Fill with random data the elements corresponding to the five throws of each friend.

3. Find the total points for each friend.

4. Find for each friend how many times the result of the throw was 6.

5. Display the results.

A program that implements the desired steps has the following steps:

1. Import the library **random** to generate the throws:

```
import random
```

2. The list is initialized as an empty array and the minimum and maximum values of the throws are given:

```
throws = [ ]
min = 1
max = 6
```

3. The array is filled up for the throws:

```
for r in range (3) :
    throws.append([ ])
    for t in range (5):
        throws[r].append(random.randint(min, max))
```

4. The array of throws is displayed:

```
for k in range(3):
    print(throws[k])
```

5. The variable for the sum is initialized and the sum is computed:

```
for r in range (3):
    throws[r].append(0)
    for t in range (5):
        throws[r][5] = throws[r][5] + throws[r][t]
```

6. The sixth column (column No. 5 `throws[5]` has the sum of the die throws for each friend.

7. To count the number of 6's for each friend, initialize the count to 0 and use a for loop.

```
friends = ['John', 'Robert', 'Alfred']
count = [ 0, 0, 0 ] # Initialize the count vector.
for r in range (3):
```

```
for t in range(5):
    if throws[r][t] == 6: # Each throw is compared to 6.
        count[r] += 1 # Increment the count for each 6.
```

8. Display the throws for each friend and the sum of points.

The number of throws with a six and the sum are displayed. There are three cases for the number of 6's. None at all, a single 6, and several 6's. Also, the friends vector with names is initialized:

```
for r in range (3):
    if(count[r] == 1):
    elif(count[r] >= 2):
    elif(count[r] == 0):
        print("The friend ", r + 1, " DID NOT have a six. " )
    print("The sum of points is: ", throws[r][5], ".")
    print()
```

9. Display the Names, throws, total, No. of 6's.

```
# Names, throws, total, No. of 6's.
for i in range(len(throws)):
    print(friends[i], '\t',throws[i], ' \t',count[i])
```

The complete final script is:

```
"' This is script Example_4_16.py
This program gives the throws of a die by three friends.
Computes the sum of the throws and finds the number of 6's
each friend threw."'

# import the library random:
import random

# Initialize an empty array and the
# minimum and maximum values of the throws:

throws = [ ]
min = 1
max = 6

# Script continues at next page.
```

```
# Script continued from previous page.

# Fill the arrays with random integers:
for r in range (3):
   throws.append([ ])
   for t in range(5):
      throws[r].append(random.randint(min, max))

# Display the array:
for k in range(3):
   print(throws[k])

# Initialize the variable for the sum and compute it:
for r in range(3):
   throws[r].append(0)
   for t in range(5):
      throws[r][5] = throws[r][5] + throws[r][t]

# Initialize count to 0 for the count
of 6's and the vector friends:
friends = ['John', 'Robert', 'Alfred']
count = [0, 0, 0] # Initialize the vector count.
for r in range (3):
   for t in range(5):
      if throws[r][t] == 6: # Each throw is compared to 6.
         count[r] += 1  # Increment the count for each 6.

# Display the throws for each friend
# and the sum of points and 6's:
for r in range (3):
   if(count[r] == 1):
      print("The friend ", r + 1, " had 6 ", count[r], " time." )
   elif(count[r] >= 2):
      print("The friend ", r + 1, " had 6 ", count[r], " times." )
   elif(count[r] == 0):
      print("The friend ", r + 1, " DID NOT have a 6. " )
   print("The sum of points is: ", throws[r][5], ".")
   print()

# Names, throws, total, No. of 6's.
for i in range(len(throws)):
   print(friends[i], '\t',throws[i], ' \t',count[i])
```

The result for a run of the program is:

```
[4, 2, 4, 2, 1]
[6, 3, 4, 2, 3]
[5, 4, 4, 6, 3]

The friend 1 DID NOT have a six.
The sum of points is: 13 .

The friend 2 had a six 1 time.
The sum of points is: 18 .

The friend 3 had a six 1 time.
The sum of points is: 22 .

# Names, throws, total, No. of 6's.
John      [4, 2, 4, 2, 1, 13]   0
Robert    [6, 3, 4, 2, 3, 18]   1
Alfred    [5, 4, 4, 6, 3, 22]   1
```

Example 4.17 System of simultaneous linear differential equations
Any linear differential equation can be written as a system of simultaneous
first-order linear differential equations. These systems can be written in matrix
form and thus, the techniques for matrices can be applied to solve them. The
procedure is shown with an example.

Let the system of simultaneous linear first-order differential equations:

$$\frac{dx}{dt} = -4x + y + z$$

$$\frac{dy}{dt} = x + 5y - z$$

$$\frac{dz}{dt} = y - 3z$$

This system of equations can be written in matrix form as:

$$\begin{bmatrix} \dfrac{dx}{dt} \\ \dfrac{dy}{dt} \\ \dfrac{dz}{dt} \end{bmatrix} = \begin{bmatrix} -4 & 1 & 1 \\ 1 & 5 & -1 \\ 0 & 1 & -3 \end{bmatrix} \begin{bmatrix} x \\ y \\ z \end{bmatrix}$$

In matrix form this equation is:

$$\frac{dX}{dt} = AX$$

where

$$A = \begin{bmatrix} -4 & 1 & 1 \\ 1 & 5 & -1 \\ 0 & 1 & -3 \end{bmatrix} \qquad \text{and} \qquad X = \begin{bmatrix} x \\ y \\ z \end{bmatrix}$$

The solution for this matrix differential equation is given in terms of the eigenvalues λ_1, λ_2, and λ_3 and the eigenvectors V_1, V_2, and V_3 of matrix A as:

$$X = C_1 V_1 e^{\lambda_1 t} + C_2 V_2 e^{\lambda_2 t} + C_3 V_3 e^{\lambda_3 t}$$

The eigenvectors are given by

$$V_1 = \begin{bmatrix} v_{11} \\ v_{12} \\ v_{13} \end{bmatrix}, \qquad V_2 = \begin{bmatrix} v_{21} \\ v_{22} \\ v_{23} \end{bmatrix}, \qquad V_3 = \begin{bmatrix} v_{31} \\ v_{32} \\ v_{33} \end{bmatrix}$$

The eigenvalues and the eigenvectors can be found with

```
import numpy as np
eigenvalues, eigenvectors = np.linalg.eig(A)
print(eigenvalues)
print(eigenvectors)
```

For the system matrix A above, the eigenvalues are given by:

```
[ 5. -4. -3.]
```

and the eigenvectors are:

```
[[-1.23091491e-01 9.90147543e-01 7.07106781e-01]
 [-9.84731928e-01 -9.90147543e-02 2.15425046e-16]
 [-1.23091491e-01 9.90147543e-02 7.07106781e-01]]
```

The eigenvectors are the columns of the array. Thus, the eigenvectors are given by:

```
V0 = eigenvectors[:, 0]
```

```
V1 = eigenvectors[:, 1]
V2 = eigenvectors[:, 2]
```

which in this example are:

```
V0 = [-0.12309149, -0.98473193, -0.12309149]
V1 = [ 0.99014754, -0.09901475, 0.09901475]
V2 = [7.07106781e-01, 2.15425046e-16, 7.07106781e-01]
```

The last eigenvector can be written as:

```
V2 = [0.707106781, 0, 0.707106781]
```

Then, the solution of the differential equations is:

$$X = C_1 \begin{bmatrix} -0.12309149 \\ -0.98473193 \\ -0.12309149 \end{bmatrix} e^{5t} + C_2 \begin{bmatrix} 0.99014754 \\ -0.09901475 \\ 0.09901475 \end{bmatrix} e^{-4t}$$

$$+ C_3 \begin{bmatrix} 0.707106781 \\ 0 \\ 0.707106781 \end{bmatrix} e^{-3t}$$

4.10   Arrays in Pandas

Pandas is a package designed to handle arrays. It is similar to numpy but it has found applications in data science and in this book it is used in the machine learning chapters. Pandas is open-source and free to use. It can be installed using the pip tool. Pandas makes use of many numpy methods and it can be said that without numpy Pandas would not exist, however, Pandas methods are much more powerful than the corresponding methods in numpy alone. It can perform high-level methods as can be seen in the examples provided. There are two main objects in Pandas, series and data frames. A series uses the keyword Series and it is a one-dimensional object. Data frames are two-dimensional objects and use the keyword DataFrame. Both objects use indexes to refer to the columns. In data frames, both columns and rows have indexes. To start using Pandas in a Python script it is necessary to import it as

```
import pandas as pd
```

A Pandas series can be created with

```
>>> a = pd.Series(['a', 'b', 1, 2])
>>> a
```

The result is

```
0       a
1       b
2       1
3       2
dtype: object
```

The Pandas series is displayed and also shows the type of the result as an object. The first column corresponds to the index and the second one to the series values. Now consider the series b and the result given by

```
>>> b = pd.Series([7, 8, 9, 0, 1, 2])
>>> b

0       7
1       8
2       9
3       0
4       1
5       2
dtype: int64
```

which indicates that the elements in the series have the type int64; that is, signed integers with a 64-bit lenght. The type of a Pandas series can be changed using the method **astype**. For example, to change the type of series b to a floating point series and to a string series use

```
>>> c = pd.astype(float)
>>> d = pd.astype(str)
>>> c
>>> d
```

The result for the series c is

```
0       7.0
1       8.0
2       9.0
3       0.0
4       1.0
5       2.0
dtype: float64
```

and for the series d

```
0       7
1       8
2       9
3       0
4       1
5       2
dtype: object
```

An index in the series can be found with `index in series` where the result is a boolean, for example,

```
>>> 2 in b
True
>>> 7 in b
False
```

By default, indexes are numbered as 0, 1, 2, 3, ..., but they can be reindexed with the method `reindex`, as for example, in the series b the indexes are changed to 1, 5, 3, 0, 2, 4 with

```
>>> b = b.reindex(index = [1, 5, 3, 0, 2, 4])
>>> b
```

```
1       8
5       2
3       0
0       7
2       9
4       1
```

The index in a series can be defined in the series definition as

```
>>> dates = pd.Series([1, 2, 2024], index = ['month', 'day', 'year'])
>>> dates
```

```
month       1
day         2
year     2024
dtype: int64
```

A name can be given to the series as

```
>>> dates.name = 'Birth'
>>> dates

month       1
day         2
year     2024
Name: Birth, dtype: int64
```

An index can be searched within a series with the method `in`. For example

```
>>> 'month' in dates
True

>>> 'Animal' in dates
False
```

Now consider changing the indexes to 'year', 'quantity', and 'weight'. This can be done with

```
>>> datesNew = pd.Series(dates, index = ['year','quantity', 'weight'])
>>> datesNew

year        2024.0
quantity       NaN
weight         NaN
Name: Birth, dtype: float64
```

The new indexes correspond to a value NaN which means Not-a-Number because there is no value assigned there. The index column may also have a name with the method `index.name`. For example,

```
>>> dates.index.name = 'Several'
>>> dates

Several
month       1
day         2
year     2024
dtype: int64
```

4.10.1 Data Frames

A data frame is an array ordered with indexes by row and by column. The keyword for a data frame is DataFrame. A very common way to generate a data frame is using a dictionary. A dictionary which contains the names of six American countries, their code, and their size in million of square km is

```
>>> data= { 'Countries': ['USA','Canada','Mexico','Brazil', \
    'Argentina','Colombia'],'Code': ['US','CA', 'MX', 'BR', 'AR', 'CO'], \
    'Size(Sq-Km)': [9.53, 9.98, 1.96, 8.51, 2.78, 1.14]}
```

A data frame is created with this dictionary

```
>>> df = pd.DataFrame(data)
>>> df
```

And the result is

```
     Countries   Code    Size
0          USA    US     9.53
1       Canada    CA     9.98
2       Mexico    MX     1.96
3       Brazil    BR     8.51
4    Argentina    AR     2.78
5     Colombia    CO     1.14
```

A data frame has a heading which in this example is retrieved with

```
>>> df.columns
```

```
Index(['Countries', 'Code', 'Size'], dtype = 'object')
```

A column can be displayed with the name. For example, to display the column corresponding to the codes

```
>>> df['Code']
```

```
0      US
1      CA
2      MX
3      BR
4      AR
5      CO
Name: Code, dtype: object
```

Alternatively, to display the same information use

```
>>> df.Code
```

To see the first five lines in the data frame use

```
>>> df.head()
```

```
       Countries   Code    Size
0            USA    US     9.53
1         Canada    CA     9.98
2         Mexico    MX     1.96
3         Brazil    BR     8.51
4      Argentina    AR     2.78
```

But if only a certain number of rows is to be displayed, use

```
>>> df.head(3)
```

```
       Countries   Code    Size
0            USA    US     9.53
1         Canada    CA     9.98
2         Mexico    MX     1.96
```

To display a row, use the index and the method `loc` as

```
>>> df.loc[2]
```

```
Countries    Mexico
Code         MX
Size         1.96
Name: 2, dtype: object
```

To display it as a row, use double square brackets as

```
>>> df.loc[[2]]
```

```
       Countries   Code    Size
2         Mexico    MX     1.96
```

To add a capital city for `Colombia` add a new column with the name `Capital` as

```
>>> df['Capital'] = df.Countries == 'Colombia'
>>> df
```

	Countries	Code	Size	Capital
0	USA	US	9.53	False
1	Canada	CA	9.98	False
2	Mexico	MX	1.96	False
3	Brazil	BR	8.51	False
4	Argentina	AR	2.78	False
5	Colombia	CO	1.14	True

The result is False for every country but Colombia because it was declared a capital city for this country. Another way to add a column is by declaring the new column with a function, for example, in data frame df add a column with half the population. This new column is called Half and it is added with

```
>>> df['Half'] = df['Size']*0.5
>>> df
```

```
>>> df['Capital'] = df.Countries == 'Colombia'  >>> df
```

	Countries	Code	Size	Capital	Half
0	USA	US	9.53	False	4.765
1	Canada	CA	9.98	False	4.990
2	Mexico	MX	1.96	False	0.980
3	Brazil	BR	8.51	False	4.255
4	Argentina	AR	2.78	False	1.390
5	Colombia	CO	1.14	True	0.570

To delete a column, use the method drop as

```
>>> df.drop('Capital', axis = 1)
>>> df
```

	Countries	Code	Size	Half
0	USA	US	9.53	4.765
1	Canada	CA	9.98	4.990
2	Mexico	MX	1.96	0.980
3	Brazil	BR	8.51	4.255
4	Argentina	AR	2.78	1.390
5	Colombia	CO	1.14	0.570

It can be seen that the column corresponding to Capital has been deleted or dropped. To delete a row just write the column index; for example, to delete the row corresponding to Canada use

```
>>> df.drop(1)
>>> df
```

```
       Countries    Code    Size      Half
0            USA      US     9.53     4.765
2         Mexico      MX     1.96     0.980
3         Brazil      BR     8.51     4.255
4      Argentina      AR     2.78     1.390
5       Colombia      CO     1.14     0.570
```

4.10.2   Generation of Data Frames

Data frames can be generated either by importing data from dictionaries as it was done in the previous subsection or from csv-files, text files, and excel files, to name a few of the ways to generate data. To import data it is necessary to have the name of the file and the path. To read csv-files the Pandas method read_csv is used. The format is

```
data = pd.read_csv('path/file name', sep = "")
```

where **sep** refers to the separator in the data. To find out the path use

```
>>> import os
>>> os.getcwd()
```

If a table A is in text format the separator may be a tab \n. The table can be read with

```
df_A = pd.read_table("A.txt", sep = "\t", names = ["Column names"])
```

As an example, to read the file 'countriesOfTheWorld.csv' in the format csv with the instruction read_csv, use

```
df = pd.read_csv('countriesOfTheWorld.csv')
```

and the result is

	Country	Region	...	Industry	Service
0	Afghanistan	ASIA	...	0,24	0,38
1	Albania	EUROPE	...	0,188	0,579
2	Algeria	AFRICA	...	0,6	0,298
3	American Samoa	OCEANIA	...	NaN	NaN
4	Andorra	EUROPE	...	NaN	NaN
..	...	...	...	...	...
222	West Bank	MIDDLE EAST	...	0,28	0,63
223	Western Sahara	AFRICA	...	NaN	0,4
224	Yemen	MIDDLE EAST	...	0,472	0,393
225	Zambia	AFRICA	...	0,29	0,489
226	Zimbabwe	AFRICA	...	0,243	0,579
	[227 rows × 20 columns]				

In this file, now converted to a Pandas data frame, there is a heading which contains a number of characteristics besides the country name. The listing ends with file information which in this case is a data frame with 227 rows and 20 columns. Due to width limitations, only a few columns and a few rows are displayed by Python. If the heading is of interest, it can be printed out with

```
print(df.head(0))
```

4.10.3   Functions for Series and Data Frames

Functions can be applied to series and data frames. As an example, consider the series a = pd.Series([1 , 4, -7, 8, 9, 10, 3, -2]). Another series can be created with elements that are greater than 2; thus

```
>>> a = pd.Series([1, 4, -7, 8, 9, 10, 3, -2])
>>> a1 = a[a > 2]
```

which produces

```
0     4
1     8
2     9
2    10
1     3
dtype: int64
```

To compute the sine function of a Pandas series, the numpy function sin can be used as

```
>>> b = pd.Series([1, 2, 3, 4])
>>> np.sin(b)
```

```
0     0.841471
1     0.909297
2     0.141120
0    -0.756802
dtype: float64
```

which correspond to sin(1.0), sin(2.0), sin(3.0), and sin(4.0), respectively. The mean, median, and standard deviation can be obtained with

```
>>> a = pd.Series([10, 4, -7, 5, 8, 9, -1, 3, -2])
>>> a.mean()
3.2222222222222223
>>> a.median()
4.0
>>> a.std()
5.65194165260439
```

Further information about a series or data frame can be obtained with the method `describe` as

```
>>> a.describe()
count      9.000000
mean       3.222222
std        5.651942
min       -7.000000
25%       -1.000000
50%        4.000000
75%        8.000000
max       10.000000
dtype: float64
```

In the case of data frames, the use of the method `describe` produces the same results for each column as shown for the following data frame

```
>>> aa = pd.DataFrame([[1, 2], [3, 5]])
>>> aa.describe()

              0         1
count   2.000000   2.00000
mean    2.000000   3.50000
std     1.414214   2.12132
min     1.000000   2.00000
25%     1.500000   2.75000
50%     2.000000   3.50000
75%     2.500000   4.25000
max     3.000000   5.00000
```

To obtain further information for a series or data frame use

```
>>> aa.info()

<class 'pandas.core.frame.DataFrame'>
RangeIndex: 2 entries, 0 to 1
Data columns (total 2 columns):
```

```
 #    Column   Non-Null Count   Dtype
 --   ----     ------------     ----
 0    0        2  non-null      int64
 1    1        2  non-null      int64
dtypes: int64(2)
memory usage: 160.0 bytes
```

The methods `loc` and `iloc` fetch information from a series or data frame. They work as:

loc - gets rows and/or columns with particular label.

iloc -gets rows and/or columns with integer locations.

To show how they work consider the data frame with indexes given by literals and not by numbers. The data frame is defined by an array 6×4 with column indexes with the letters a, b, c, d, e, and f. The row indexes are w, x, y, and z. The data frame is

```
>>> df = pd.DataFrame(np.arange(24).reshape(6,4), \
        index=list('abcdef'), columns=['w', 'x', 'y', 'z')
>>> df
```

```
     w    x    y    z
a    0    1    2    3
b    4    5    6    7
c    8    9   10   11
d   12   13   14   15
e   16   17   18   19
f   20   21   22   23
```

The function `loc` retrieves by indicating the indexes. For example, to retrieve the element in row d and column y, use

```
>>> df.loc['d', 'y']
14
```

To obtain row 'd' use

```
>>> df.loc['d']
d   12   13   14   15
```

To retrieve rows from row a to row c use

```
>>> df.loc['a': 'c']

    w   x   y   z
a   0   1   2   3
b   4   5   6   7
c   8   9   10  11
```

The method `iloc` is similar to `loc` but it requires to give the indexes as integers. For example, to display the rows from row `a`, which is row 1, to row `d`, which is row 3, use `df.iloc[1: 4]` as

```
>>> df.iloc[1: 4]
b   4   5   6   7
c   8   9   10  11
d   12  13  14  15
```

To obtain a slice of the data frame use both row and column indexes as in

```
>>> df.iloc[1:4,1:3]
b   5   6
c   9   10
d   13  14
```

In Chapters 11 and 12 corresponding to machine learning, arrays with Pandas format are greatly used and it is seen that they make the treatment of arrays easier.

4.11   Python Instructions for Chapter 4

Table 4.3 shows the instructions used in Chapter 4.

TABLE 4.3:   Python instructions for Chapter 4

Instruction	Description
astype	Method to change the type of a Pandas object.
append	Used to add elements to an array.
array	Conversion of a matrix or list to an array.
comprehension	Simplified way to create an array.
DataFrame	A two-dimensional Pandas object.
describe	Method to get statistical information for an object.
dot	Dot product of vectors or arrays.
drop	Pandas method to delete a row/column.

(Continues next page.)

TABLE 4.3: Python instructions in Chapter 4 (*Continued*)

Instruction	Description
eye	Generates an identity array.
getcwd	Method to find out the working directory.
eye	Method to read rows in a data frame.
in	Method to search in a Pandas object.
eye	Method to find out information for the object.
loc	Method to search in a Pandas object.
matrix	Conversion of an array or list to a matrix.
pandas	Module to handle arrays.
read_csv	Method to read CSV files.
read_table	Method to read tables.
reindex	Change the indexing in a Pandas object.
series	Define a series in Pandas.
sort	Used to numerically sort an array.
transpose	Obtains the transpose of an array.

4.12 Conclusions

Vectors, matrices, and arrays find applications in many areas of science, mathematics, finance, engineering, and in general everyday life. This chapter has introduced the use of arrays with Python. A section describes the use of Pandas to handle arrays. Several examples have presented how to use Python, numpy and Pandas functions.

4.13 Selected Bibliography

1. A hybrid algorithm, https://numerentur.org/ordenacion-de-vectores-timsort/. Consulted on March 19, 2024.

2. Tim Peters, https://en.wikipedia.org/wiki/Tim_Peters_(software_engineer). Consulted on March 19, 2024.

4.14 Exercises

1. Given the lists a = [1, 4, -2, 9], b = [-9, 0, 4, 7, -8]. Compute a + b, a - b and 3*a.

2. For the vector a = [23, 27, 02, 58, 19, -18, -1, 32, 92]. Find:

    (a) a[3]
    (b) a[3:5]
    (c) a[:2]
    (d) a[0:2]

3. The magnitude or Euclidean norm of a vector $x = [x_1, x_2, ..., x_n]$ is defined by:

$$|x| = ||x|| = \sqrt{x_1^2 + x_2^2 + ... + x_n^2}$$

Write a program to compute the magnitude for the vector x = [2, -4, 6, -3, -9].

4. The infinite norm $||x||_\infty$ of a vector is defined by the largest magnitude element. For the vector in the previous exercise, the infinite norm is equal to 9. Write a program that obtains the infinite norm of a vector.

5. The p norm is defined by:

$$||x||_p = (x_1^p + x_2^p + ... + x_n^p)^{1/p}$$

Compute the norms 3 and 4 for the vectors in Exercise 1.

6. The scalar product of two vectors x = $[x_1, x_2, ..., x_n]$, $y = [y_1, y_2, ..., y_n]$, also called dot product, is defined by:

$$x \cdot y = x_1 \cdot y_1 + x_2 \cdot y_2 + ... + x_n \cdot y_n$$

Write a script to obtain the dot product of two vectors and compare the result using the **np.dot** function.

7. The scalar product is also defined as the product of the vector magnitudes times the cosine of the angle between them. In this:

$$X \cdot Y = |X| \cdot |Y| \cos(\theta)$$

Figure 4.1 shows the angle for two vectors. Now, find the angle for vectors X = [3, 4, 5], X = [-2, -1, 6].

8. Generate a random vector of size 10 with integral elements and sort it. Display both vectors.

9. The data for a list of football players has:

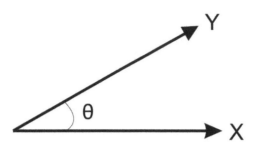

FIGURE 4.1: Angle between two vectors.

- String with the player's name
- Age
- Height
- Weight
- Position

Write a program that must read data for 10 players. The program must select those players that

- Are older than a certain user given age,
- Are higher than a certain user given height,
- Are heavier than a user given weight,
- Play a user given position.

Use arrays for the solution.

10. For matrices A y B shown, compute $AB = A * B$ y $BA = B*A$.

$$A = \begin{bmatrix} 1 & 0 & 2 \\ 2 & 3 & -1 \\ -1 & 0 & 3 \end{bmatrix}$$

$$B = \begin{bmatrix} 1 & 2 & 3 \\ 7 & 4 & -2 \\ 6 & -10 & -5 \end{bmatrix}$$

11. For the matrices in the previous exercise, compute $A + B$ and $A - B$.

12. For matrix A, which of matrices B or C is the inverse of A.

$$A = \begin{bmatrix} 4 & 8 & 2 \\ 2 & -2 & -1 \\ 6 & 7 & 3 \end{bmatrix}$$

$$B = \begin{bmatrix} -1 & 10 & 4 \\ 12 & 0 & -8 \\ -26 & 20 & 24 \end{bmatrix}$$

$$C = \begin{bmatrix} -1 & 10 & 4 \\ 12 & 0 & -8 \\ -26 & -20 & 24 \end{bmatrix}$$

13. For the matrices in the previous exercise compute AB y CA. Explain the results.

14. Compute the product $A * b$.

$$A = \begin{bmatrix} 2 & 1 & 0 \\ 9 & -4 & 11 \\ 16 & 5 & -8 \end{bmatrix}$$

$$b = \begin{bmatrix} 10 \\ -3 \\ 17 \end{bmatrix}$$

15. Find the transpose of A for the previous exercise.

16. The game of memory, also called concentration game, consists of placing pairs of cards upside down and arranged in matrix form. The players take turns to choose two cards. If the cards are equal the cards are assigned to that player. If the cards are not equal they are placed upside down again. The game ends when all pairs are uncovered. Use a 5×2 matrix of cards.

17. Improve the memory game to allow three levels: rookie with a 2×5 matrix, intermediate with a 4×4, and pro with 4×6 matrix.

18. A diagonal matrix has non-zero elements in the main diagonal and has 0 otherwise; that is, it is of the form:

$$D = \begin{bmatrix} d_{0,0} & 0 & 0 & \cdots & 0 \\ 0 & d_{1,1} & 0 & \cdots & 0 \\ 0 & 0 & d_{2,2} & \cdots & 0 \\ \vdots & & & & \\ 0 & 0 & 0 & \cdots & d_{n-1,n-1} \end{bmatrix}$$

Propose an algorithm that generates a diagonal matrix with random integers.

19. A triangular matrix has 0s either above or below the main diagonal. If the triangular matrix has all the elements below the main diagonal equal to zero the matrix is called upper triangular. Similarly, the matrix is called lower triangular if the elements above the main diagonal are equal to zero. An upper triangular matrix is of the form:

$$U = \begin{bmatrix} a_{0,0} & a_{0,1} & a_{0,2} & \cdots & a_{0,n-1} \\ 0 & a_{1,1} & a_{1,2} & \cdots & a_{1,n-1} \\ 0 & 0 & a_{2,2} & \cdots & a_{2,n-1} \\ \vdots & & & & \\ 0 & 0 & 0 & \cdots & a_{m-1,n-1} \end{bmatrix}$$

And a lower triangular matrix is of the form:

$$L = \begin{bmatrix} a_{0,0} & 0 & 0 & \cdots & 0 \\ a_{1,0} & a_{1,1} & 0 & \cdots & 0 \\ a_{2,0} & a_{2,1} & a_{2,2} & \cdots & 0 \\ \vdots & & & & \\ a_{m-2,0} & a_{m-2,1} & a_{m-2,2} & \cdots & 0 \\ a_{m-1,0} & a_{m-1,1} & a_{m-1,2} & \cdots & a_{m-1,n-1} \end{bmatrix}$$

Write a script to generate triangular matrices. It must ask if the matrix is going to be upper or lower triangular.

20. The trace of an array is given by the sum of the elements of the main diagonal. The elements of the main diagonal have equal indices; that is, they are of the form.

A[k][k]

Then, the trace is given by:

$$A = \sum_{k=0}^{n-1} A[k][k]$$

Implement an algorithm that computes the trace of a matrix. Then compare using the function **numpy.trace()**.

Chapter 5

Functions

5.1 Introduction

So far in the book, small scripts o programs have been developed. In real life, a program can have thousands of lines of code. In such scripts, it might be possible to repeat code in different parts. By grouping that code in subprograms, and when needed just call these subprograms, a great deal of coding could be saved, making the program shorter and easier to understand and debug. In other cases, the script is too long and it may be partitioned into smaller subprograms to make it easier to understand and maintain. In a larger program, often referred to as the main program, a subprogram is *called* by the main program. In this chapter, techniques to develop subprograms are described. Several examples using subprograms are presented in the chapter. Also presented is the concept of recursion which is the case when a subprogram is used by itself.

5.2 Subprograms

A subprogram is a part of a larger program or script. They may also have inputs and outputs. A subprogram is usually called a function. Examples of functions available in Python are the trigonometric functions, the square root, the functions available in `numpy`, `math`, etc.

There are two types of subprograms, namely, procedures and functions. *Procedures* are subprograms that have inputs but that do not provide an output. Rather they execute some tasks, for example, print out data. *Functions* instead receive input data and provide output data to be used in other functions or in the main program. Some programmers and authors refer to procedures and functions as simply functions. In object-oriented programming, functions are called methods and they belong to a class. Thus, they may be referred to as either functions or methods.

DOI: 10.1201/9781003222118-5

5.3   Functions in Python

There are two kinds of functions in Python. Those defined and available in modules or libraries, such as `sin(x)` available in the library `math`, and those defined by the user. This chapter covers user defined functions. The format for a function in Python is:

```
def function_name(parameters):
        body of the function
        return output_data
```

The keyword **def** indicates that a function is defined in the following instructions which form the body of the function. In the case of procedures, there is not a **return** statement. This is used to return to the main program the variables of interest. As an example, if a function is to obtain the sum of the first N integers, that function can be:

```
def adder(N):
        sum = N*(N+1)/2
        return sum
```

The function **adder** receives from the main program the value of N, and the function returns the **sum**. This is received by the main program for whatever is needed.

Functions must be defined before using them in another function or in the main program. If this is not done an error is produced. There are some guidelines that must be followed to write functions:

- The name of the function is unique and it cannot be used by any other function or variable.

- The variables between the parenthesis after the function name are called arguments or parameters. The number is usually fixed, but in some cases is variable (see section 5.9).

- The names of the parameters can change.

- The variables defined in a function are not the same as the variables in another function or in the main program.

- The body of the function must be indented. The first row is not indented after the body of the function is not part of the function.

Example 5.1 Prime numbers

The algorithm to find out if a given number is a prime number from Example 2.5 can be rewritten using functions. This can be done in the following way. The important part of the algorithm can be implemented in a function as:

```
''' This is script Example_5_1a.py
  This function finds if an integer is a prime number.'''

def prime(number):
   # Given a positive integer, determine if it is a prime.

   # Import necessary functions:
   from math import sqrt
   is_prime = True # Suppose it is a prime number.
   divisor = 2 # Start with 2 until
           # sqrt(number) unless an exact divisor is found.
   while divisor <= sqrt(number) and is_prime:
      quotient = number//divisor # Integer division.
      if quotient * divisor = num:
          is_prime = False # It is not a prime number.
      else:
          divisor = divisor + 1 # Continue searching.
   return is_prime
# End of the function.
```

The function can be called from a main programs as

```
   prime(number)
```

The main program follows:

```
''' This is script Example_5_1b.py
This program calls the function prime to find whether a number
is a prime.'''

   # Insert function here.

n = int(input("Enter an integer number: "))
is_prime = prime(n)  # The function prime is called here.
      if (is_prime):    # prime must be True.
         print("It is a prime number." )
      else:
         print("It is NOT a prime number." )
```

The function must be placed before it is called. The script including the main program and the function is

```
"' This is script Example_5_1c.py
This program calls the function prime to find if a number is a
prime.'"

# Beginning the function.
def prime(number):
    # Given a positive integer, determine if it is a prime.

    # Import necessary functions:
    from math import sqrt
    is_prime = True # Suppose it is a prime number.
    divisor = 2 # Start with 2 up to sqrt(number)
            # unless an exact divisor is found.
    while divisor <= sqrt(number) and is_prime:
       quotient = number//divisor # Integer division.
       if quotient * divisor == number:
           is_prime = False # It is not a prime number.
       else:
           divisor = divisor + 1  # Continue searching.
    return is_prime
# End of the function.

n = int(input("Enter an integer number: "))
for num in range(n):
    is_prime = prime(n) # The function primeis called here.
    if (is_prime):       # prime must be True.
      print("It is a prime number.")
    else:
        print( "It is NOT a prime number.")
```

A run for n = 20 and n = 29 gives the result:

```
Enter an integer number: 20
It is NOT a prime number.

Enter an integer number: 29
It is a prime number.
```

Example 5.2 Solution of a second degree equation

To solve a second-degree equation $ax^2 + bx + c = 0$ it is necessary to calculate the square root of the discriminant $d = \sqrt{b^2 - 4ac}$ which is done in a function. The values of a, b and c are read in the main program. The function to calculate the square root of the discriminant takes as input the coefficients of the second-order equation and returns the square root of the discriminant. This function is:

```
''' This is script Example_5_2.py

This function computes the square root of the discriminant.'''

def discriminant_root(a1, b1, c1):
    d = math.sqrt(b1**2 - 4*a1*c1)
    return d
```

The main program is:

```
''' This is script Example_5_2a.py

This is the main program.'''

# Main program
import math

# Enter coefficients.
a = float(input("Enter the coefficient for x**2: "))
b = float(input("Enter the coefficient for x:  "))
c = input("Enter the independent term coefficient:  ")

# Compute the square root of the discriminant.
d1 = discriminant_root(a, b, c)

# Compute the roots:
x1 = (-b + d1)/2/a
x2 = (-b - d1)/2/a

# Display the roots:
print("The first root is: ", x1)
print("The second root is: ", x2)
```

This algorithm can be further partitioned. Two additional functions can be defined, the first one reads the coefficients and the second one computes and displays the roots. This makes the main program more compact and easy to understand. The functions are:

```
def read_coeff( ):
   # This function reads the coefficients of a second degree
equation.

   # Read the coefficients a, b, c.

   a = float(input("Enter the value of a: "))
   b = float(input("Enter the value of b: "))
   c = float(input("Enter the value of c: "))

   return [a, b, c] # The coefficients are in a list.
```

The function **read_coeff** does not receive arguments but it must include the empty parentheses. It returns to the main program the three coefficients in a list. The coefficients are passed then from the main program to the function **discriminant_root**.

The second function computes the roots x1 and x2:

```
def compute(a, b, c, d):

   # This function receives the coefficients
   # and the square root of the discriminant.

   x1 = (-b + d)/2/a
   x2 = (-b - d)/2/a

   return (x1, x2) # Returns x1 and x2 in a tuple.
```

The main program is then:

```
# Main program

"" This program computes the roots of a second degree equation
of the form: a*x**2 + b*x + c = 0.""

import math
# Read the coefficients:
[a, b, c] = read_coeffs( )

# Compute the square root of the discriminant:
d1 = discriminant(a, b, c)

# Script continues at next page.
```

```
# Script continued from previous page.

# Compute the roots:
(x1, x2) = compute(a, b, c, d1)

# Display the roots:
print("The first root is: ", x1)
print("The second root is: ", x2)

# End of the main program
```

A run for the equation first calls the function `read_coeff` to obtain the coefficients, then calls the function to compute the square root of the discriminant, and finally computes the roots. If the discriminant is zero, then the square root cannot be computed and the program marks an error. For the equation:

$$2x^2 + 3x - 2 = 0$$

The roots obtained are:

x1 = -2 and x2 = 0.5.

In the final complete script, the functions must be placed before they are called. In this example then the function `read_coeff` is the first one followed by the functions `discriminant_root` and `compute`. Finally, the main program is the last one in the final script.

5.4   Recursion

Functions defined so far are executed in the order they are called. In some other cases, it is convenient to use a technique known as *recursion*. In this technique, a function calls itself; that is, within the function, there is a call to itself. Such functions are called *recursive* functions.

As with any technique, there are advantages and disadvantages in using recursion. Some advantages are:

- The written code is more elegant and clean.

- A complex task can be made simpler.

- It may preclude the use of nested iterations.

On the other side, some disadvantages may be:

- The logic may be harder to follow.

- Recursive calls are usually slower and take longer to execute.

- Debugging may be more difficult.

- There must be a condition to stop the recursion to avoid infinite loops.

Recursion is shown by means of an example.

Example 5.3 Factorial using recursion
The function to compute the factorial can be written as:

```
def factorial_Rec(n):
    fact = 1 # Initialize factorial variable to unity.
    if (n > 1):
        fact = n*factorial_Rec(n - 1) # Factorial is computed
                                      # recursively.
    else:
        fact = 1
    return fact # Returns the variable fact.
```

It can be seen that the function `factorial` calls itself at:

```
fact = n*factorial_Rec(n - 1)
```

A run with **n = 6** is

```
>>> factorial_Rec(6)
```

which gives the result

```
720
```

as was expected. It can also be called from a main program.

5.5   Anonymous Functions or `lambda` Functions

Sometimes a function is as short as a single line. In this case, the function can be defined as an anonymous function using the keyword `lambda` instead of the `def` keyword. An example is the function discriminant that can be written in a single line as `math.sqrt(b1**2 - 4*a1*c1)`. The format for the `lambda` function is:

```
lambda arguments: expression
```

For example, for the discriminant function, the `lambda` form expression is:

```
d = lambda a1, b1, c1: math.sqrt(b1**2 - 4*a1*c1)
```

and it can be used as:

```
import math
d = lambda a1, b1, c1: math.sqrt(b1**2 - 4*a1*c1)
print('The square root of the discriminant is:', d(1,9, 20))
```

and the result is:

```
The square root of the discriminant is: 1.0
```

which is the correct value as expected.

5.6   Pass by Reference

When a variable is defined in Python, the program assigns a reference to it. This reference is maintained during the execution of the program. For a variable defined in a function, a reference is defined too. For example, in the program:

```
a = 23
print('Outside the function a = ', a)
def test(a):
    a = 27
    print('Inside the function: a = ', a)
print('Outside the function: a = ', a)
```

The result is

```
Outside the function: a = 23
Inside the function: a = 27
Outside the function: a = 23
```

The first **print** occurs outside the function and the value of a is 23, then the function, is called and the variable **a** is defined and printed and the result is 27. Finally, the last print is outside the function and the value is 23. The reference to the variable **a** outside the function is different from the variable **a** inside the function. For the main program, there is no change in the value of **a**.

In the case of a list is the same unless the list is somewhat modified by, for example, adding or deleting elements. For example,

```
a = [23, 32]
print('Outside the function a = ', a)
def test(a):
    print('Inside the function: a = ', a)
    a.append(54)
    print('Inside the function: a = ', a)
test(a)
print('Outside the function: a = ', a)
```

The result is

```
Outside the function: a = [23, 32]
Inside the function: a = [23, 32]
Inside the function: a = [23, 32, 54]
Outside the function: a =  [23, 32, 54]
```

The value is modified inside the function but here is referenced to the list defined before calling the function and thus it is the same list but modified. In the last **print** instruction, the list is already modified.

5.7   Local and Global Variables

The variables defined in a main program can be used by a function when passed as arguments. The variables defined inside a function can only be used in that function and do not have any reference outside the function. They are called local variables.

A variable can be made global inside the functions that use the global variable and this makes the variable available to all other functions and the main program. To make a global variable only has to be indicated as:

```
global name_of_variable
```

As an example consider the script:

```
a = 24
print('Outside the function: a = ', a)

def test():
    global a
    print('Inside the function: a = ', a)
    a = 54
    print('Inside the function: a = ', a)
```

```
def test2():
    global a
    print('Inside the function test2: a = ', a)
    a = 71

test()
test2()
print('Outside the function: a = ', a)
```

The result is:

```
Outside the function a: = 24
Inside the function: a = 24
Inside the function: a = 54
Inside the function test2: a = 54
Outside the function: a = 71
```

The variable a is defined as `global` in the functions `test` and `test2`. The values are modified in the functions and this is seen in the main program where the last `print` produces the last value modified in the function `test2`.

5.8   Keyword and Default Arguments

Up to this point, the arguments of a function have been specified by its position and the value. That is, in the definition:

```
def function(var1, var2):
```

the arguments `var1` and `var2` always have the same position and when the function is called, the first value is assigned to `var1` and the second value is assigned to `var2`. When the function is called as in:

```
x = function(3, -8)
```

The values assigned to `var1` and `var2` are 3 and -8, respectively.

When keyword arguments are used, the call to the function identifies the arguments by the name, regardless of the position in the function call. For the example above, the call can be made with a keyword argument if it is made as either:

```
x = function(var1 = 3, var2 = -8)
```

or as:

```
x = function(var2 = -8, var1 = 3)
```

Another variation for keyword arguments is when not all of the variables are associated with a name. In that case, the position may be important. For example:

```
x = function(var2 = -8, 3)
```

In this case the variable var2 = 2 and var1 = 3.

Example 5.4 Keyword arguments
When dealing with the function discriminant from Example 5.2, it can be verified that for a = 1, b = 3, and c = 2, the result is d = 1. Running this function with keyword arguments in the following script:

```
"' This is script Example_5_4.py
This script shows the use of keyword arguments.'"
import math

# Define the function.
# The arguments are not in the same order as
#   in Example 5.2.
def discriminant_root( b1, c1, a1 = 2):
     d = math.sqrt(abs(b1**2 - 4*a1*c1))
     return d

b1 = 5

# The call to the function does not include
#   a1 which in the function definition is equal to 2.
d = discriminant_root( b1, c1 = 2)
print('The square root of the discriminant is ', d)
b = 5
c = 2
d = discriminant_root( b, c)
print('The square root of the discriminant is ', d)
```

The result is

```
The square root of the discriminant is 3.0
The square root of the discriminant is 2.0
```

which are the correct results.

5.9   Variable-length Arguments

Sometimes the list of arguments can be different when calling a function. Then, it is useful to have variable-length arguments. These are arguments that they may not be present in the calling of the function. Those arguments are preceded by an asterisk.

Example 5.5 Variable-length arguments
A function is to receive two numbers. The numbers are summed if a variable add is present. If it is not present the numbers are subtracted. Then the function is:

```
"' This is script Example_5_5.py
This script shows the use of variable-length arguments.'"

# This function has a variable-length argument list.
# If the variable add is present the other two arguments
# are added. Otherwise, they are subtracted.
def addSubtract(a, b, *add):
    if add:
        return a + b
    else:
        return a - b

# Values of a and b:
a = 6
b = 4

# First call to the function with add = 1:
d1 = addSubtract(a, b, 1)
print(d1)

# Second call to the function. add is omitted:
d1 = addSubtract(a, b)
print(d1)
```

The result of the first call to the function is the addition $6 + 4$ because add $= 1$. In the second call add is not present and the result is the subtraction 6 - 4:

10

2

5.10   Additional Examples

Example 5.6 Throw of a die

When a die is thrown the result is random. To simulate the die the function **random** can be used. This function is included in the library **random**. The format for this function is:

```
import random
x = random.random( )
```

This generates a random number between 0 and 1. Another function for random numbers is:

```
x = random.randint(a, b)
```

which generates a random integer number between **a** and **b**. For example, the following script generates random integers between 10 and 76:

```
import random
for k in range(10):
    x = random.random( )*(76 - 10) + 10
    print(x)
```

and to obtain 10 integers between 10 and 76, a **for** loop is used as:

```
import random
for i in range(10):
    x = random.randint(10, 76)

    print(x)
```

For this example, the number has to be between 1 and 6. The script is:

```
def throw_die( )
    from random import randint
    number = randint(1, 6)
    print( 'The number is: ', number)
```

This function is run as:

```
>>> throw_die( )
The number is: 5

>>> throw_die()
The number is: 2
```

The number obtained is always between 1 and 6.

Example 5.7 Fibonacci series

In a Fibonacci series, an element of the series is obtained as the sum of the two previous elements. The first and second elements of the series, are 0 and 1, respectively. The following elements can be obtained with the equation:

$$\text{fibonacci}(n) = \text{fibonacci}(n - 1) + \text{fibonacci}(n - 2)$$

A function that computes the elements of the Fibonacci series is:

```python
def fibonacci(m):
    # The first two elements:
    list_fibo = [0, 1]

    # Elements from 3 to m + 1:
    for k in range(2, m + 1)
        list_fibo.append(f[k - 2] + f[k - 1])

    # The list is displayed:
    print( 'The Fibonacci series is:', list_fibo)
```

This function produces the first m elements in the Fibonacci series. This function does not have a **return** because the **print** instruction is in the function. This type of functions are often known as procedures. A run for $m = 10$ is:

```python
>>> fibonacci(10)

[ 0, 1, 1, 2, 3, 5, 8, 13, 21, 34 ]
```

Example 5.8 Sum and product of the digits of a number

It is desired to look for integer numbers whose sum of its digits is equal to the product of the digits, for example, in the number 123 the sum and the product of the digits is equal because $1 + 2 + 3 = 1 \times 2 \times 3$. To look for these types of numbers, four functions are needed:

- The first function extracts the digits from the number.

- The second function receives the digits and returns the sum.

- The third function receives the digits and returns their product.

- The fourth and last function compares both results and displays the result.

The algorithm to obtain the digits is to divide the number by 10 and the important part is the residue which gives the digits. The residue can be obtained with the function `modulus`. For example, for the number 239, dividing by 10 the residue is 9 which is the least significant digit, or the rightmost digit of the number. This number corresponds to the units. The quotient is 23 which is divided by 10 again to obtain the residue 3 which is the next digit to the left. This digit corresponds to the tens and the quotient is 2. This is the last digit corresponding to the hundreds. This digit is called the most significant digit. Now that the three digits are obtained the sum is 14 and the product is 54 and the number 239 does not meet the requirements.

A function that extracts the digits from a given integer number is:

```
"' This is script Example_5_8.py
This function extracts the digits of an integer number."'
# Function to obtain the digits and reverse them.
def digits(number):
# Start the process with a while loop:
# Define an index and initialize the list digits:
      digits = [ ]
      index = number

# Start the while loop:
      while index >= 10
            quotient, residue = divmod(index, 10)
            index = quotient
            # Appends the residue to the list digits:
            digits.append(residue)
            # Append the last digit to the list:
      digits.append(index)
            # Reverse the list to have the digits
            # in the correct order:
      digits.reverse()

      return digits
```

After finding the digits the next task is to find the sum and the product of them. To find the sum a **for** loop is used. The argument for the function is the list containing the digits. The function to add the digits is:

```
"' This is script Example_5_8C.py
This function adds the digits of an integer number."'

def addition(list_digits):
      # Initialize the sum to 0:
      add = 0
      # Start the for loop to add the digits:
      for k in range(len(list_digits)):
            add = add + list_digit[k]
      return add
```

The function to obtain the product of the digits uses another `for` loop and it is:

```
"' This is script Example_5_8C.py
This function multiplies the digits of an integer number.'"

def multiply(list_digits):
    # Initialize the product:
    product = 1
    # Start the for loop to find the product of the digits:
    for k in range(len(list_digits)):
        product = product*list_digit[k]
    return product
```

The last function is to compare the product and the sum. If they are equal then print a message of success. The function is:

```
"' This is script Example_5_8C.py
This function compares the addition and the product of the
digits of an integer number.'"

def compare(number, add, product):
    # In a condition the sum and the product are compared:
    if add == product:
        print('The digits of ', number, 'satisfy the criterion.')
```

The complete script is:

```
"' This is script Example_5_8C.py
  This program finds integer numbers such that
  the sum and product of their digits are equal.'"

def find_digits(number):
    # Start the process with a while loop:
    # Define an index
    digits = [ ]

# Script continues at next page.
```

```
# Script continued from previous page.

      index = number
      while index >= 10
            quotient, residue = divmod(index, 10)
            index = quotient
            digits.append(residue)
      digits.append(index)
      digits.reverse()
      return digits

def addition(list_digits):
      add = 0
      for k in range(len(list_digits)):
            add = add + list_digit[k]
      return add

def multiply(list_digits):
      product = 1
      for k in range(len(list_digits)):
            product = product*list_digit[k]
      return product

def compare(number, add, product):
      if add == product:
            print('The digits satisfy the criterion.')
      else:
            print('The digits do not satisfy the criterion.')

# Main program
# Read in the number
for number in range(0, 1000):
      digits = find_digits(number)
      add = addition(digits)
      product = multiply(digits)
      compare(number, add, product)
```

For the first 1000 integers, the following numbers satisfy the condition: 22, 123, 132, 213, 231.

Example 5.9 Perimeter and area of a triangle

Given the coordinates of the vertices of a triangle, it is desired to obtain the perimeter and the area. The program must include a function to compute the perimeter and an other function to compute the area. Another function reads in the coordinates of the vertices.

The function to read the coordinates is named `read_vertices()`. The coordinates of each vertex are stored in a list. The function is:

```
"' This is script Example_5_9.py
This function reads the coordinates of a vertex."'

def read_vertices(i):
        # This function reads the vertices coordinates.
        # Define the list:
        p = [ ]

# Read the coordinates and separate them in a list of strings:
        print('Enter the x,y-coordinates for vertex', i, \
                'separated by a comma:')
        p = (input( ))
        p = split(',')

# Change the type to float:
        p = [float(p[0]), float(p[1])]
# Return the list of the two coordinates:
        return p # A vertex coordinates are returned.
```

To compute the distance between two vertex the Euclidean distance formula is used. For two point $P_1 = (x_1, y_1)$ and $P_2 = (x_2, y_2)$ the distance is given by:

$$distance = \sqrt{(x_1^2 - x_2^2) + (y_1^2 - y_2^2)}$$

The coordinates received are a list of strings; thus, they have to be converted to `float` type. The function is:

```
"' This is script Example_5_9C.py
This function finds the distance between two vertices."'

def dist(p1, p2):
        # Distance between two vertices is computed.
        # The distance is computed using
        # the Euclidean distance formula.
        # Each coordinate is converted to a
        # float type before computing:
        distance = sqrt((p1[0] - p2[0])**2+(p1[1] - p2[1])**2)
        return distance
```

The perimeter is given by the sum of the sides computed with the function dist and the area is given by the equations:

$$S = perimeter/2$$
$$Area = \sqrt{S(S - L_1)(S - L_3)(S - L_3)}$$

where L_1, L_2, and L_3 are the lengths of the triangle sides and S is an intermediate variable. Then the function to calculate the perimeter and the area is:

```
"' This is script Example_5_9C.py
This function finds the perimeter and area of a triangle.'"

def perimeter_area(A, B, C):
        # This function computes the three distances
        # between vertices, the perimeter, and the area.
        L1 = dist(A,B)
        L2 = dist(B,C)
        L3 = dist(C,A)

# Computation of the perimeter:
        perimeter = L1 + L2 + L3

# computation of the area:
        S = perimeter/2
        area = sqrt(S*(S - L1)*(S - L2)*(S - L3))

# Display of perimeter and area:
        print('The perimeter is: ', perimeter)
        print('The area is: ', area)
```

The main program is:

```
"' This is script Example_5_9C.py
This program finds the perimeter and area of a triangle.'"
# Main program
# Read vertices:
A = read_vertices(1)
B = read_vertices(2)
C = read_vertices(3)
perimeter_area(A, B, C)
```

A run for this program is:

```
Enter the x, y-coordinates for vertex 1 separated by a comma:
0, 0
```

```
Enter the x, y-coordinates for vertex 2 separated by a comma:
0, 1
Enter the x, y-coordinates for vertex 3 separated by a comma:
1, 0
The perimeter is: 3.414213562373095
The area is: 0.49999999999999983
```

The correct result for the area is 0.5 and for the perimeter is $2 + \sqrt{2} = 3.414213562373095$. Thus, the values obtained are correct.

5.11 Python Instructions in Chapter 5

Table 5.1 shows the instructions used in Chapter 5.

TABLE 5.1: Python instructions for Chapter 5

Instruction	Description
def	It is used to define a function.
return	Indicate the variables that the function returns.

5.12 Conclusions

Functions are implementations of subalgorithms. They allow programmers to partition and simplify a large program which otherwise might be too cumbersome to implement. Functions may also be used to contain code that is repeated several times in a large program. The chapter contains a number of examples that allow the reader to become proficient in the use of functions.

5.13 Exercises

1. Consider the following Python script. Show and explain the results that the script produces at each stage:

```
def mistery(n) :
    t = 0
    for i in range (1, 4):
        m = n * i
```

```
        print(n, " X ",  i,  " are ", m)
        t = t + m
    return(t)


# Main program.
    num = int(input("Number ")
    print("The result is:  ", mistery(num))
```

2. It is desired to write an algorithm to generate multiplication tables. The script should have a function that reads the integer n which indicates the desired table. Another function must obtain the corresponding table. Finally, a last function must print out the table. The table must be obtained from n×1 up to n×10.

3. Using lambda functions, write algorithms to obtain the multiplication tables for n = 7 and n = 11.

4. Generate a matrix or array with three rows and two columns. The procedure must be the following:

   (a) With a function initialize the matrix with 0's.

   (b) In another function, change the 0's by random integer numbers.

   (c) In another function obtain the transpose matrix.

   (d) Finally, in another function display both matrices, the original and the transpose ones.

5. Repete the previous exercise using global variables.

6. Goldback's conjecture states that any even number is formed by the sum of two prime numbers. Write a Python script to check the conjecture. It must comply with the following:

   (a) Read a number in a function.

   (b) In a second function check if the number is an even one.

   (c) If the result in the second function is affirmative, in a third function find two prime numbers that satisfy Goldback's conjecture

   (d) Finally, in a fourth function display the two prime numbers. If the number given is not an even one, the fourth function must display a message saying that the number given is not an even number.

Examples are: n = 16 = 3 + 13, n = 38 = 1 + 37, n = 120 = 3 + 117.

7. The table to find the income tax is shown below. Implement an algorithm that a person must pay. The Python script must consist of a function that receives the income. Another function must evaluate the tax amount, and a third function must display both amounts, the income and the tax amounts.

Lower limit	Upper limit	Fixed tax	Tax rate
0	999	0	5%
1,000	9,999	50	10%
10,000	99,999	100	15%
100,000	999,999	10,000	25%
1,000,000		100,000	35%

For example, if a person has a yearly income of 234,000, the tax that must be paid is:

100,000 + 0.25*234,000 = 68,500

8. In a convenience store, the Promotion of the day is the following: After paying for the purchase, the customer chooses a number between 1 and 5. If the number is the same as the last number in the ticket purchase number, the customer receives a 5% discount.

Write an algorithm using a function to generate the random number between 1 and 5. In a second function compare with the ticket number that is also generated with a random function but in this case, the ticket number must consist of 5 digits. In a third function compare both numbers and notify if whether or not a discount is obtained. The input data read by the main program is the purchase amount.

9. A procedure to find out the date for Easter Sunday is given below:

   (a) Obtain a, b, c, and d as:

   $a = year\%19$
   $b = year\%4$
   $c = year\%7$
   $d = (19 * a + 24)\%30$
   $e = (2 * b + 4 * c + 6 * d + 5)\%7$

   (b) The required date is obtained with: march 22 + d + e

   Design an algorithm that evaluates this date. The algorithm must read the year and it must print out the date for Easter Sunday. Given the fact that the date can correspond to April, consider this fact to print out the correct date.

Chapter 6

Object-Oriented Programming

In this chapter, a brief description is given on how to write programs using an object-oriented paradigm with Python. Demonstrative examples are also presented.

6.1 Introduction

This chapter is a brief introduction to Object-Oriented Programming (OOP) as it is implemented in Python. Although this way of programming is more associated with languages such as C++ and Java, among others, it has also been incorporated into Python; thus making it a multiparadigm programming language. A great deal of textbooks have been dedicated to the teaching of OOP; thus this section only presents the main characteristics of OOP in Python and gives the specifics to start programming using this paradigm. The chapter begins by defining the elements of OOP and continues with examples of how to program in Python emphasizing the OOP.

6.2 The Object-Oriented Programming Paradigm

Object-oriented programming became popular in the 80s and has had a boom since 1990. It is used in high-level languages such as C++ and Java. One advantage of using OOP is that computational objects can be used to represent real-life objects in a very clear way. In this way OOP can be used to solve problems more clearly and easily, allowing the design of more efficient algorithms.

Object-oriented programming requires a clear and precise understanding of the following terms used in OOP and defined below:

- object
- attribute

- method

- class and instance

- encapsulation

- inheritance

- polymorphism

An **object** is a key element in OOP. It can represent a real object such as a bank account, a client or a transaction. Each object has a state and a behavior. The state of an object is the set of characteristics that define the object at a given instant of time. The behavior is described by the activities associated with the object.

Attributes or **properties** are the values that are stored internally by the object and that can be primitive data or other objects. The attributes or properties represent the state of the object.

A **method** is a set of programming instructions which are given a name. When a method is invoked, the program instructions are executed. The methods of an object define its behavior, for example, the actions that can be performed on it.

A **class** is the model or template from which an object is created. In this way, the important features of the object are defined in both its state (or structure) and its behavior (set of methods). From a class, several different objects with similar characteristics can be defined. A specific example of a class is called an **instance** which in turn is called an object. Instances of the same class have the same type of attributes or characteristics and share the methods of the class.

To create **instances** of a class, it is necessary to include a constructor method in the class definition, which will receive a variable number of arguments to specify the initial conditions of the properties of the instance and return the constructed object with designated values.

To **encapsulate** means that information can be stored and that also can be worked with it so that status changes can only be made by the methods of the object and that no other object can change them. Encapsulation allows the hiding of the internal details of the implementation of these methods.

Some classes relate to each other through **inheritance**, forming a classification of hierarchy. Inheritance allows the definition and creation of new objects such as specialized types of preexisting objects. Objects inherit the properties and behavior of all classes to which they belong. New objects can share (and extend) their characteristics (attributes) and behavior (methods) without having to re-implement them.

Polymorphism is the property that allows OOP values of different types of data to be handled using a uniform interface. The same method can be used by several classes, but differently. When the same method is applied to different objects, although they share the same name of the method, the behavior will be produced for the type of object that is invoking it. When this occurs during "runtime", this feature is called *late assignment* (late binding) or dynamic allocation.

6.3 Classes in Python

To create a class, in a new script, the first line is the keyword `class` followed by the name of the class name which must start with a capital letter. Then the attributes and the methods of the class must be defined. The attributes are defined and initialized in a constructor.

As an example consider the class that includes countries and their capital cities:

```
class Countries:

  #Class methods:

  #Constructor
  def __init__(self, country, capital):
    self.country = country
    self.capital = capital

  # This is a method to print out the result.
  def output(self):
    print('The capital of '+ self.country + 'is ' + self.capital)
```

The class is stored in the file `classCountries.py`. In this class, the attributes are `country` and `capital`. The methods are `__init__`, which is the constructor where the attributes are defined and initialized, and `output` which prints out the result. The word `self` refers to the object created in the instantiation. Thus, when the methods appear `self.country` and `self.capital`, they refer to the attributes `country` and `capital` of the object or instance of the class. An example run is as follows:

```
>>> a = Countries('France', 'Paris')
>>> b = Countries('Canada', 'Ottawa')
>>> c = Countries('USA', 'Washington DC')
```

The objects are a, b, and c and they are *instances* of the class `Countries`. The information for any instance can be printed out with:

```
>>> a.output()
The capital of France is Paris
```

As it can be seen, following the object-oriented paradigm in Python, data and actions are combined into a single object called *class*. In the class definition, data is called `attributes` and actions are called `methods`. This file should include a `constructor` which is responsible for creating new objects. The constructor is a method that allows the initialization of the properties of an object. Every class must have a constructor. In this example, the constructor method assigns initial values to the attributes `country` and `capital` of the instances created from the class `Countries`. The constructor keyword is `__init__` (two underscores before and after `init`). For the class `Countries`, the constructor initializes the values of the attributes.

In the definition of the properties of classes, Python handles primitive data that can be of different types such as numeric, symbolic or string type and may also include instances of other classes.

6.3.1   Creation and Use of a Class

Let us suppose that it is needed to calculate a fraction which consists of a numerator and a denominator. To avoid a division by zero, the divisor has to be checked to be different from zero. Then a new class called `MyFraction1` is defined, for which the numerator and denominator are declared as their attributes. As it has been said before, methods are defined by functions as shown below:

```python
class MyFraction1:
    # This class obtains the ratio of two numbers.

    #Class methods

        # Constructor
        def __init__(self, numerator, denominator):
            self.numerator = numerator
            if denominator == 0:
                print('The denominator must be different from 0.')
            else:
                self.denominator = denominator

        # This method evaluates the fraction and calls the method
        # to print out. The method result evaluates the fraction:
        def result(self):
            self.DisplayResult(self.numerator/self.denominator)

        # The method DisplayResult prints out the result
        # of the fraction:
        def DisplayResult(self, number):
            print('result = ', number)
```

If this class is now used to obtain the fraction, we have to create the object invoking an instance of the class `MyFraction1` as follows:

```
>>> a = MyFraction1(2, 3)
>>> a.result()
result = 0.6666666666666666
```

6.3.2 Declaration and Use of Setters and Getters

The attributes declared in the class `MyFraction1` are global attributes and can be changed from outside the class. Continuing with the object defined above as `a`, the numerator can be changed to 1 as:

```
>>> a.numerator = 1
```

To avoid this modification of the attributes from outside the class, encapsulation can be used. To modify the contents, methods called `setters` can be added to set the values of the numerator and denominator. This is done with the following methods:

```
def setNumerator(self, numerator):
    self.numerator = numerator

def setDenominator(self, denominator):
    if denominator == 0:
        print('The denominator must be different from 0.')
    else:
        self.denominator = denominator
```

The complete class, renamed `MyFraction2`, is:

```
class MyFraction2:
    # This class obtains the ratio of two numbers.
    # It includes setters.

    #Class methods

        # Constructor
        def __init__(self, numerator, denominator):
            self.numerator = numerator

# Script continues next page.
```

```
# Script continued from previous page.

        if denominator == 0:
            print('The denominator must be different from 0.')
        else:
            self.denominator = denominator

    # Setters
    def setNumerator(self, numerator):
        self.numerator = numerator
    def setDenominator(self, denominator):
        if denominator == 0:
            print('The denominator must be different from 0.')
        else:
            self.denominator = denominator

    # The method result evaluates the fraction:
    def result(self):
        self.DisplayResult(self.numerator/self.denominator)

    # The method DisplayResult prints out
    # the result of the fraction:
    def DisplayResult(self, number):
        print('result = ', number)
```

The methods `setNumerator` and `setDenominator` are known as `setters` since they allow the initialization (set) of variables. The initialization of the values of the numerator and denominator can be done in either way as shown in the following example for the setting of `numerator` and `denominator`. Thus:

```
>>> a = MyFraction2(2, 3)
>>> a.setNumerator(7)
>>> a.denominator = 21
>>> a.result()

0.3333333333333333
```

Now, another instance of the class is created as follows:

```
>>> b = MyFraction2(2, 3);
>>> b.setNumerator(1)
>>> b.setDenominator(4)
>>> result = b.result()
```

```
0.25
```

Object b is another instance of the class MyFraction2 and the values of its properties are different and independent of those belonging to object a.

The getters are methods to retrieve an attribute. Getters for this class can be as simple as:

```
# Getters
def getNumerator(self):
    return self.numerator

def getDenominator(self):
    return self.denominator
```

Once the getters are inserted into the class MyFraction2 they can be used as:

```
>>> a.getDenominator()
>>> a.getNumerator()
```

and the result is the numerator and denominator for the values defined for the object a.

6.3.3 Static Methods

The class MyFraction2 can be improved by printing out the result as a fraction as, for example, 2/3. This can be done by adding an additional print instruction in the method DisplayResult as

```
def DisplayResult(self, number):
    print('result = ', number)
    print('Fraction = ', int(self.num), '/', int(self.denom))
```

In addition, the fraction can be reduced to its minimum parts as in

$$7/21 \rightarrow 1/3$$

In order to do this, the greatest common divisor (gcd) of the numerator and denominator has to be found and then they have to be divided by it. The gcd can be found with the method gcd from numpy and it only works for integer positive numbers and thus, the numbers have to be converted to integer arguments. A method to do this is:

```
def reduce(num, denom):
   import numpy as np
   num = int(num)
   denom = int(denom)

   # Evaluate greatest common divisor:
   mygcd = np.gcd(np.abs(num), np.abs(denom))

   # Numbers are reduced:
   return (num/mygcd, denom/mygcd)
```

This method can be added to the class as a static method simply adding the decorator @staticmethod on top of the method. A static method is a method bounded to the class and not to the objects. It can modify a class attribute or parameter. This method can be called each time an attribute is read. Thus, it has to be called from the constructor and from the setter. The new class is renamed MyFraction3 as:

```
class MyFraction3:
   # This class obtains the ratio of two numbers.
   # It displays the fraction using the GCD.

#Class methods

   # Constructor
   def __init__(self, numerator, denominator):
      self.numerator = numerator
      if denominator == 0:
         print('The denominator must be different from 0.')
      else:
         self.denominator = denominator
         (self.numerator, self.denominator) = \
            MyFraction3.reduce(self.numerator, self.denominator)

   # Setters
   def setNumerator(self, numerator):
      self.numerator = numerator
      (self.numerator, self.denominator) = \
         MyFraction3.reduce(self.numerator, self.denominator)

# Script continues at next page.
```

```
# Script continued from previous page.

  def setDenominator(self, denominator):
    if denominator == 0:
      print('The denominator must be different from 0.')
    else:
      self.denominator = denominator
      (self.numerator, self.denominator) = \
      MyFraction3.reduce(self.numerator, self.denominator)
  # Getters
  def getNumerator(self):
    return self._numerator

  def getDenominator(self):
    return self.denominator

  @staticmethod
  def reduce(num, denom):
    import numpy as np

    # Numbers are made integer type:
    num = int(num)
    denom = int(denom)
    # Evaluate greatest common divisor:
    mygcd = np.gcd(np.abs(num), np.abs(denom))

    # Numbers are reduced:
    return(num/mygcd, denom/mygcd)

  # The method result evaluates the fraction:
  def result(self):
    self.DisplayResult(self.numerator/self.denominator)

  # The method DisplayResult prints out
  # the result of the fraction:
  def DisplayResult(self, number):
    print('result = ', number)
```

As an example, the fraction 9/81 is reduced with:

```
>>> a = MyFraction3(9, 81)
>>> a.result()
result = 0.111111111111111
Fraction = 1 / 9
```

6.3.4   Encapsulation

Encapsulation is a fundamental concept in OOP. Encapsulation of a class is
the packing of data and methods that restrict or protect them from access by
external methods. Such encapsulated methods or attributes are called private.
To accomplish this in Python, a protected name attribute or method has to
be prefixed by a single underscore. For example, in the class `MyFraction`, the
numerator can be made private just by prefixing it with an underscore, as in:

```
def __init__(self, numerator, denominator):
    self._numerator = numerator
                        ⋮
```

This prefixing has to be made in every reference to `self._numerator`. The
class is changed to (The attribute `self._numerator` is shown in bold type):

```
class MyFraction4:
  # This class obtains the ratio of two numbers.
  # It displays the fraction using the GCD.

#Class methods

  # Constructor
  def __init__(self, numerator, denominator):
    self._numerator = numerator
    if denominator == 0:
      print('The denominator must be different from 0.')
    else:
      self.denominator = denominator
      (self._numerator, self.denominator) = \
      MyFraction4.reduce(self._numerator, self.denominator)

  # Setters
  def setNumerator(self, numerator):
    self._numerator = numerator
    (self._numerator, self.denominator) = \
    MyFraction4.reduce(self._numerator, self.denominator)

# Script continues at next page.
```

```
# Script continued from previous page.

  def setDenominator(self, denominator):
    if denominator == 0:
      print('The denominator must be different from 0.')
    else:
      self.denominator = denominator
      (self._numerator, self.denominator) = \
      MyFraction4.reduce(self._numerator, self.denominator)

  @staticmethod
  def reduce(num, denom):
    import numpy as np

    # Evaluate greatest common divisor:
    mygcd = np.gcd(np.abs(num), np.abs(denom))

    # Numbers are reduced:
    return (num/mygcd, denom/mygcd)

  # The method result evaluates the fraction:
  def result(self):
    self.DisplayResult(self._numerator/self.denominator)

  # The method DisplayResult prints out
  # the result of the fraction:
  def DisplayResult(self, number):
    print('result = ', number)
    print('Fraction = ', int(self._num), '/', int(self.denom))
```

An example run is:

```
>>> b = MyFraction4(2, 3)
>>> b.result()
result = 0.6666666666666666
Fraction = 2 / 3
```

The numerator is changed to 8:

```
>>> b.numerator(8)
Traceback (most recent call last):
File "<pyshell#56>", line 1, in <module>
b.numerator(8)
AttributeError: 'MyFraction4' object has no attribute 'numerator'
```

Using the setter:

```
>>> b.setNumerator(8)
>>> b.result()
result = 2.6666666666666665
Fraction = 8 / 3
```

Methods can also be made private in the same way.

6.3.5   Inheritance

The properties of classes can be inherited from existing classes. Thus, the class that inherits is called a subclass or a child class. A top-level class is sometimes called a superclass or a parent class. Each subclass can alter the properties which were inherited and add also their own. A subclass inherits all data and behavior of the parent class. It can also add more attributes and more methods of its own.

In the last definition of the class of the example, the class `MyFraction` is a superclass for any subclass defined from it.

As an example, a subclass must obtain the reciprocal of the fraction given. The subclass must receive the numerator and denominator from the superclass. Then, the definition of the subclass must include the parent class as:

```
class Reciprocal(MyFraction):
```

The class `Reciprocal` is made a subclass of `MyFraction` by enclosing the name of the parent class in parentheses after the name of the child class as is shown above. Now the parameters of the child class must be declared as convenient starting with *self*. Since the new fraction is to be inverted, the parameters are declared in the reverse order in the constructor as:

```
def __init__(self, denom, num):
```

In addition, in order to use every attribute available in the parent class, the parent class constructor call must be included prefixed with the word `super().`, and for this subclass is

```
super().__init__(num, denom)
```

Finally, the method from the parent class used in the subclass is the method `result` which is prefixed with the word `self`, and is included in the constructor as

```
self.result()
```

The complete subclass is:

```
class Reciprocal(MyFraction):

    def __init__(self, denom, num):
        super().__init__(num, denom)
        self.result()
```

and the complete class, renamed **MyFraction5**, is:

```
class MyFraction5:
    # This class obtains the ratio of two numbers.
    # It displays the fraction using the GCD.

# Class methods

    # Constructor
    def __init__(self, numerator, denominator):
        self._numerator = numerator
        if denominator == 0:
            print('The denominator must be different from 0.')
        else:
            self.denominator = denominator
            (self._numerator, self.denominator) = \
            MyFraction5.reduce(self._numerator, self.denominator)

    # Setters
    def setNumerator(self, numerator):
        self._numerator = numerator
        (self._numerator, self.denominator) = \
        MyFraction5.reduce(self._numerator, self.denominator)

    def setDenominator(self, denominator):
        if denominator == 0:
            print('The denominator must be different from 0.')
        else:
            self.denominator = denominator
            (self._numerator, self.denominator) = \
            MyFraction5.reduce(self._numerator, self.denominator)

# Script continues at next page.
```

```
# Script continued from previous page.

  @staticmethod
  def reduce(num, denom):
    import numpy as np

    # Evaluate greatest common divisor:
    mygcd = np.gcd(np.abs(num), np.abs(denom))

    # Numbers are reduced:
    return (num/mygcd, denom/mygcd)

  # The method result evaluates the fraction:
  def result(self):
    self.DisplayResult(self._numerator/self.denominator)

  # The method DisplayResult prints out
  # the result of the fraction:
  def DisplayResult(self, number):
    print('result = ', number)
    print('Fraction = ', int(self._num), '/', int(self.denom))

class Reciprocal(MyFraction5):

  def __init__(self, denom, num):
    super().__init__(num, denom)
    self.result()
```

As an example, if the numerator and denominator are 2 and 3, respectively, then run MyFraction5 from either the file menu Run or by pressing F5, and then execute:

```
>>> c = Reciprocal(2,3)
result = 1.5
Fraction = 3 / 2
```

which is the reciprocal of the fraction given.

6.3.6   Overloading

Overloading is the mechanism that allows a method to accept different types of arguments and to have an already existing name. When trying to add two objects a and b

```
>>> a = MyFraction5(3, 7)
>>> b = MyFraction5(2, 5)
```

the result is:

```
>>> a + b
Traceback (most recent call last):
 File "<pyshell#98>", line 1, in <module>
    a + b
TypeError: unsupported operand type(s) for +: 'MyFraction5'
  and 'MyFraction5'
```

Python reports an error because the addition of this type of object has not been defined. Recall that for the basic arithmetic operations of addition (+), subtraction (-), multiplication (*) and division (/), Python establishes the equivalence with the following methods: add, sub, mul and truediv, for the addition, subtraction, multiplication, and division, respectively. The addition of two fractions is given by

$$\frac{num1}{denom1} + \frac{num2}{denom2} = \frac{num1*denom2 + num2*denom1}{denom1*denom2}$$

where num and denom represent the numerator and denominator of each fraction. The method __add__ is overloaded as

```
def __add__(self, other):
```

The parameters for the addition of two fractions are the objects previously defined.

The numerator and denominator of the resulting fraction are, respectively:

```
self.num = self.num*other.denom + self.denom*other.num
self.denom = self.denom*other.denom
```

The resulting fraction is sent to the method **reduce** to simplify the fraction:

```
self.num, self.denom = self.reduce(self.num, self.denom)
```

The resulting numerator and denominator are sent to the method **result** to display the resulting fraction:

```
self.result()
```

The complete method is given now:

```python
def __add__(self, other):
   self.num = self.num*other.denom + self.denom*other.num
   self.denom = self.denom*other.denom
   self.num, self.denom=self.reduce(self.num, self.denom)
   self.result()
```

In the same way, a method can be implemented for the subtraction of the two fractions by rewriting the plus sign in the numerator sub, which will change the fraction:

```python
def __sub__(self, other):
   self.num = self.num*other.denom - self.denom*other.num
   self.denom = self.denom*other.denom
   self.num, self.denom=self.reduce(self.num, self.denom)
   self.result()
```

By incorporating these definitions of new methods to perform arithmetic operations on the fractions, the class MyFraction5 is improved and renamed MyFraction6.

```python
class MyFraction6:
   # This class obtains the ratio of two numbers.
   # It displays the fraction using the GCD.

# Class methods

   # Constructor
   def __init__(self, numerator, denominator):
     self._numerator = numerator
     if denominator == 0:
       print('The denominator must be different from 0.')
     else:
       self.denominator = denominator
       (self._numerator, self.denominator) = \
       MyFraction6.reduce(self._numerator, self.denominator)

# Script continues at next page.
```

```
# Continued from previous page.

  # Setters
  def setNumerator(self, numerator):
    self._numerator = numerator
    (self._numerator, self.denominator) = \
    MyFraction6.reduce(self._numerator, self.denominator)
  def setDenominator(self, denominator):
    if denominator == 0:
      print('The denominator must be different from 0.')
    else:
      self.denominator = denominator
      (self._numerator, self.denominator) = \
      MyFraction6.reduce(self._numerator, self.denominator)

  @staticmethod
  def reduce(num, denom):
    import numpy as np
    # Evaluate greatest common divisor:
    mygcd = np.gcd(np.abs(num), np.abs(denom))
    # Numbers are reduced:
    return (num/mygcd, denom/mygcd)

  # The method result evaluates the fraction:
  def result(self):
    self.DisplayResult(self._numerator/self.denominator)

  # The method DisplayResult prints out the fraction:
  def DisplayResult(self, number):
    print("result = ", number)

  # Methods add and sub
  def __add__(self, other):
    self.num = self.num*other.denom + self.denom*other.num
    self.denom = self.denom*other.denom
    self.num, self.denom = self.reduce(self.num, self.denom)
    self.result()

  def __sub__(self, other):
    self.num = self.num*other.denom - self.denom*other.num
    self.denom = self.denom*other.denom
    self.num, self.denom = self.reduce(self.num, self.denom)
    self.result()

# Script continues at next page.
```

```
# Continued from previous page.

class Reciprocal(MyFraction6):

    def __init__(self, denom, num):
        super().__init__(num, denom)
        self.result()
```

Two instances of `MyFraction6` are created to perform the addition and subtraction operations, as shown below:

```
>>> a = MyFraction6(2, 14)
>>> b = MyFraction6(1, 4)

>>> a + b
   result = 0.39285714285714285
   Fraction = 11 / 28
```

Thus, it is observed that the result is correct. Testing for the subtraction:

```
>>> a - b
   result = -0.10714285714285714
   Fraction = -3 / 28
```

as expected.

6.4 Example

In this section, an example of classes is shown that illustrates the main features of Python for applying the object-oriented paradigm.

Example 6.1 The opening and movements of a bank account

This example shows the way to implement the movements associated with a bank account, using object orientation offered by Python. The movements that must be performed are opening an account, deposits, withdrawals, and the displaying of the balance. The main characteristic of any account is the balance; thus it is necessary to specify methods for making deposits, withdrawals and obtaining the balance. After each deposit and/or withdrawal it is also required to see the balance, for which a special method to display it will be used.

Let us begin with the declaration of the class called `Bank_Account` whose basic structure is shown in the following Script:

```
class Bank_Account:
    # It creates and manages bank accounts.
```

Class methods

The first method is the constructor that must bear the same name as the class which will create the account and assign a value to the opening balance. Each object account will then be an object or an instance of the class `BankAccount`. The constructor must read the account name and the value to be assigned to the opening balance. An appropriate constructor is:

```
# Constructor
    def __init__(self, balance):
        self.balance = balance):
```

A method to print the account balance is given the name `output` and is defined by

```
def output(self):
    print("This account has $ ", self.balance)
```

A call to this method can be incorporated in the constructor in order to display the current balance after deposits and withdrawals as:

```
# Constructor
    def __init__(self, balance):
        self.balance = balance):
        self.output()
```

Now the methods to deposit and withdraw are written. In the first case, it is an addition of the amount received to the previous balance, and in the second case, it is subtracted. Also for consideration is the case where the withdrawal is greater than the balance. In both methods, the account represents an instance of the class `BankAccount` to modify and the amount represents the amount to add or subtract from the balance of the account. For the deposit, the method to define is:

```
def deposit(self, amount)
    # Increments the balance by amount.
    self.balance = self.balance + amount)
    self.output()
```

and for the withdrawal the method is:

```
def withdrawal(self, amount):
    # Decrease account balance by the amount.
    if amount > self.balance:
        print('Insufficient funds in the account.')
    else:
        self.balance = self.balance - amount
        self.output()
```

The complete class is then:

```
class Bank_Account:
# It creates and manages bank accounts.


    Class methods

    # Constructor
    def __init__(self, balance):
       self.balance = balance):
       self.output()

    def output(self):
       print('This account has $ ', self.balance)

     def deposit(self, amount)
        # Increments the balance by amount.
        self.balance = self.balance + amount)
        self.output()

    def withdrawal(self, amount):
        # Decrease account balance by the amount.
        if amount > self.balance:
           print('Insufficient funds in the account.')
        else:
            self.balance = self.balance - amount
           self.output()
```

To test how it works, an object named Peter is defined:

```
>>> Peter = BankAccount(12)
    This account has $12.00
```

```
>>> Peter.deposit(8)
    This account has $20.00

>>> Peter.withdrawal(14);
    This account has $6.00

>>> Peter.withdrawal(8)
    Not enough funds in the account.
```

6.5 Python Instructions for Chapter 6

Instructions introduced in Chapter 6 are presented in Table 6.1.

TABLE 6.1: Python instructions for Chapter 6

Instruction	Description
init	It initializes the attributes of a class.
get	It is used to get the attributes of a class.
self	It makes reference to the same object.
set	It is used to set the attributes of a class.

6.6 Conclusions

In this chapter, an introduction to using Python to write programs using the paradigm of Object-Oriented Programming was given. Only the most basic definitions have been explored since the topic is quite large and is beyond the aims of this book and chapter. The interested reader can consult the references for a greater knowledge of the subject. However, with this chapter, it is already possible to create classes of OOP in Python.

6.7 Exercises

1. It is known that the Earth's gravity is 6 times larger than the Moon's gravity. Write a class to convert the weight on Earth to that on the Moon.

2. Write a class for the conversion from Celsius degrees to Fahrenheit and vice versa.

3. Write the methods for multiplication and division for the class `MyFraction6`.

4. Write a class with the name `Cards`. Assume that there are 52 cards in the deck. Each hand has five cards given to three players. Use random methods to give the hands and to find out which player is the winner.

5. Write a class to make the four basic operations with complex numbers.

6. Given a list of integer numbers, write a class to perform the following functions on the list:

a. Read the list.

b. Print out the list.

c. Add a new integer to the list.

d. Search for a number in the list.

e. Search how many times a number is in the list.

f. Delete a number from the list.

6.8   Selected Bibliography

[1] R. D. Neidinger, Introduction to Automatic Differentiation and MATLAB Object-Oriented Programming, SIAM Rev., Computer Science, Mathematics, vol. 52, pp. 545–563, 2010.

[2] D. Phillips, Python 3 Object-Oriented Programming: Build robust and maintainable software with object-oriented design patterns in Python 3.8, Packt Publishing, 3rd Ed., 2018.

[3] A. Downey, Think Python: How to Think Like a Computer Scientist, 2nd Ed., O'Reilly Media, 2016.

Chapter 7

Reading and Writing to Files

7.1 Introduction

So far, input data has been entered through the keyboard and the output has been displayed on the screen. Data can also be stored in a file so it can be read from this file and the output written either to the same file or to a different one. In this chapter, the different ways to read from and write data to a file are presented. Data is stored in files as bits (**binary digits**) that can have the value of 0's and 1's. Bits are grouped in bytes, which are collections of eight bits. There are mainly two types of files, namely, text files and data files. Text files consist of sequences of alphanumeric and other special symbols and they are human readable. Examples of text files are word documents. On the other hand, binary files such as executable files, images, videos, and compressed files, among others, are not human readable. In this chapter, it is shown how to store data in both, text and binary files.

7.2 Writing Data to a File

In a computer, a data file is a sequence of bytes, it also has a beginning and an end. In order to write to or to read from a file the first step is to open the file. At the end of the process, the file has to be closed. Each time a file is opened, a handle has to be generated. Python makes reference to the open file using the handle and it is used to close the file. The following instruction is used to open a file:

```
handle = open(file_name, mode)
```

The variables are:

- **open** is the keyword to open a file.

- **file_name** is the file to be opened or to be created.

DOI: 10.1201/9781003222118-7

TABLE 7.1:   Modes to handle files

Mode	Description
'r'	Opens the file for read only.
'w'	Opens the file to write. If the file exists, it deletes the existing data.
'a'	Opens the file to write. If the file exists, it appends the new data at the end of the existing data.
'r+'	Opens the file to read and write.

- **mode** is a variable which indicates the actions to be made on the file.

- **handle** is a variable used to refer to the file.

The variable **mode** has the following options shown in Table 7.1.

7.2.1   Writing Alphanumeric Data to a File

Once the file is open, the next step is to write in it. This can be done using the instruction **write** with the following format:

    handle.write('phrase')

where **phrase** is the information that is desired to write. After finishing the writing process, the file has to be closed with:

    handle.close( )

With the instruction **write** only strings can be written to a file.

Example 7.1 Writing text to a file
First, a name for the file has to be chosen. For this example the file name is **animals.txt**. A handle for the file has to be chosen. The handle is an identifier that has to be used to close the file. In the file **animals.txt**, name of animals are written:

    handle = open('animals.txt', 'w')

The kind of animals are: **cats**, **dogs**, **birds**. To write each name in a newline an escape "\n" is inserted between the names. The instruction to write is then:

```
handle.write("cats \n")
handle.write("dogs \n")
handle.write("birds \n")
```

Finally, the file is closed with:

```
handle.close( )
```

The complete script is saved to a file named `Example_7_1.py` as shown here:

```
# This is script file Example_7_1.py
# It writes animal names to a file.

handle = open('animals.txt', 'w' )
handle.write("cats \n")
handle.write("dogs \n")
handle.write("birds \n")
handle.close( )
```

After running and opening the file `animals.txt` with a text editor `animals.txt` the contents can be seen. The instruction `write` DOES NOT generate a newline automatically and, therefore, an escape sequence is included.

Example 7.2 Writing to a file with `writelines`
The previous example can be changed to use the instruction `writelines` in the following way, after opening the file as:

```
handle = open('animals.txt', 'w')
```

The `writelines` instruction is used to write all the lines in the file (in this example only a single line is available):

```
handle.writelines(["cats \n", "dogs \n", "birds \n"])
```

The file is closed with:

```
handle.close( )
```

the complete script is:

```
# This is script Example_7_2.py
# It writes names of animals with writeline.

handle = open('animals.txt', 'w')
handle.writelines(["cats \n", "dogs \n", "birds \n"])
handle.close( )
```

7.2.2 The Instruction with

The instruction with provides still another way to open a file for writing and reading. The advantage with this instruction is that the file is closed after using this instruction. The format is:

```
with open( "file_name.txt", "modo" ) as handle:

    handle.write( " phrase" )
```

The first line ends with a semicolon and the following lines have to be indented, as it was the case for conditions and loops. When the with block ends the file is closed.

Example 7.3 Writing alphanumeric data

It is desired to create a file named teams.txt with the names of the soccer national teams that have won the World Cup. The team names are: Brazil, Germany, Italy, Argentina, France, Uruguay, England, and Spain. The handle is foot and the data is written in the following instructions:

```
# This is script Example_7_3.py
# It uses writelines.

with open( "teams.txt", "w") as foot:
    foot.writelines(["Brazil\n Germany\n Italy\n Argentina\n \
    France\n Uruguay\n England\n Spain\n"])
```

The file is closed when the with statement ends. The country names appear in a column in the file teams.txt. In the script the *explicit continuation line* "\" is used to join the two lines that form the list of countries.

7.3 Writing Numerical Data to a File

If both alphanumeric and numeric data have to be written to a file, the numeric data has to be converted to string format with str( ).

Example 7.4 Writing numerical data to a file

Following up from the previous example, now it is required to create the file teams2.txt, but in addition, it is also required to write the championships won. A list with the data is given as:

```
data = ["Brazil 5\nGermany 4\n \
Italy 4\nArgentina 2\nFrance 2 \n \
Uruguay 2\nEngland 1\nSpain 1"]]
```

This list can be written to the file as:

```
# This is script Example_7_4.py

file = open("teams2.txt", "w")
data = ["Brazil 5\n Germany 4\n Italy 4\n Argentina 2\n\
    France 2\n Uruguay 2\n England 1\n Spain 1"]
file.writelines(data)
file.close( )
```

When the file teams2.txt is opened it shows the following data:

```
Brazil 5
Germany 4
Italy 4
Argentina 2
France 2
Uruguay 2
England 1
Spain 1
```

The instruction **write** can also be used to store numeric data with a format. For example, for the variable a = 3.2316 the instruction is:

```
file.write("%s"   %a)
```

Example 7.5 Alternate form for writing numerical data

Writing alphanumeric and numeric data can be done. As an example, in the baseball American league the Arizona Diamondbacks has won three pennants, after creating the file pennants.dat with

```
handle = open('pennants.dat', 'w')
```

The data is defined as pennant $= 3$ and the name of the team and the numerical data can be written as

```
handle.write("Diamondbacks \t  %s"   %pennants)
```

Finally, the file is closed

```
handle.close( )
```

The complete file is

```
# This is script Example_7_5.py
# It uses write to store data in a file.

handle = open('pennants.dat', 'w')
pennants = 3
handle.write("Diamondbacks \t  %s"    %pennants)
handle.close( )
```

In this case, the team name, a tab, and the number of pennants are written to the file.

The techniques presented so far can be used to create an array of data. In the following example, it is shown how to create this array or table.

Example 7.6 Creating a table in a file

It is desired to write the following table in a file:

1.2	3.0	-7.45
-8.3	6	17.563
98.78	-12.5	-46.2332
2	-567.2	8.43
1	0	1.2013

The table has five rows and three columns. For Python this is an array or matrix and the name given is `dataTable`:

```
dataTable = [ [ 1.2, 3.0, -7.45],
             [ -8.3, 6, 17.563],
             [ 98.7, -12.5, -46.2332],
             [ 2.1, -56.2, 8.43],
             [ 1.0, 0, 1.2013] ]
```

The script to write this table to the file `Table.dat` starts with the creation of the file `Table.dat`:

```
handle = open("Table.dat", "w")
```

Two nested loops are used to write each row, column by column, each and every element to the file Table.dat:

```
for row in dataTable:
    for column in row:
```

The instruction **write** can only have strings as arguments. Therefore, each element in the table has to be converted to a string with str(column). It is convenient to add as a separator a couple of tabs '\t'. The resulting string is named **string1** and formed and written to the file with

```
string1 = str(column) + '\t\t'
handle.write(string1)
```

This finishes the nested **for** loop. Continuing with the **for** loop for rows, it is necessary to add an end of line with the escape sequence '\n'

```
handle.write("\n")
```

Finally, the file is closed with:

```
handle.close( )
```

The complete file is:

```
# This is script Example_7_6.py
# It writes data in tabular form.

dataTable = [ [ 1.2, 3.0, -7.45] ,
              [ -8.3, 6, 17.563] ,
              [ 98.7, -12.5, -46.2332] ,
              [ 2, -56.2, 8.43],
              [ 1, 0, 1.2013] ]

handle = open("Table.dat", "w")

for row in dataTable:
    for column in row:
        string1 = str(column) + '\t\t'
        handle.write(string1)
    handle.write("\n")

handle.close( )
```

When the file Table.dat is open with the WordPad, the contents are:

1.2	3.0	-7.4
-8.3	6	17.5

98.7	-12.5	-46.2332
2.1	56.2	8.43
1	0	1.2013

7.4   Data Reading from a File

To read data from a file, the data format must be known. From the previous section, it is known that data is stored in strings. Thus, numerical data must be converted to the appropriate format. As is the case for writing to a file, the first action is to open the file for reading. This is done with:

```
handle = open("file_name", "r")
```

Then, the data is read according to the data format. This is shown with several cases.

7.4.1   Reading Data from a File

Data in a file can be either alphanumeric or numeric. In addition, data can be combined in the same file. The first example shows the case of numeric data.

Example 7.7 Reading numerical data
It is desired to read integer numbers from a file, add them and take the average. The file `integers.dat` contains the following numerical data:

```
10.4
23.7
-34.8
496.17
17.28
-23.54
```

The first step is to open the file for reading. The handle is now `my_file`:

```
my_file = open("integers.dat", 'r')
```

Next, the actual data is read each row at a time, each row contains a string that has to be converted to an integer. To sum up the integers, first initialize the sum to zero, then add up each number as is read. A count of the integers read is needed to take the average. The sum is 130 and the total number of elements is 6. Thus, the average is 21.6666666666. The last step is to close the file. First, the sum and the number of elements read are initialized to 0 .

```
sum = 0
count = 0
```

Then, the elements are read with a for loop:

```
for row in my_file:
```

The elements read from a file are strings and they have to be converted to floating-point numbers and then added to the sum. The counter count that counts the number of rows read has to be incremented by 1:

```
sum = sum + float(row)
count += 1
```

Finally, the loop is exited and the sum is divided by count to obtain the average which is also printed with two decimal digits:

```
average = sum/count
print('The average of the sum is %0.2f' %average)
```

Finally, the file is closed:

```
my_file.close( )
```

The complete file is:

```
# This is script Example_7_7.py
my_file = open("integers.dat")
sum = 0
count = 0
for row in my_file:
    sum = sum + float(row)
    count += 1
average = sum/count
print( "The average of the sum is %0.2f" %average)
my_file.close( )
```

After running the script the result is 81.53.

Example 7.8 Reading alphanumeric y numeric data

The file `temperatures.dat` contains the average monthly Celsius temperatures in 2014 in Cairo, Egypt:

Jan	13.8
Feb	15.2
Mar	17.4
Apr	21.4
May	24.7
Jun	27.3
Jul	27.9
Aug	27.9
Sep	26.3
Oct	23.7
Nov	17.1
Dec	15

It is desired to read the data from the file and display it. The information can be represented as a dictionary. The month corresponds to the `key` and the temperature to the `value`. After opening the file, each row is read and the key and value are separated with `row.split`. A `for` loop is used. The Python script starts by opening the file for reading:

```
f = open( "temperatures.dat", "r")
```

Then, the sum is initialized to 0:

```
sumT = 0
```

To read the rows a `for` loop is used:

```
for row in f:

    f.read(row)
```

The row read is split:

```
data = row.split( )
```

The first part is the month and the second one is the temperature:

```
month = data[0]
temperature = data[-1]
```

The row that was just read is displayed:

```
word = month +  " " + temperature
print (word)
```

The temperature just read is converted to a floating number and added to the accumulated sum:

```
sumT += float(temperature)
```

The average temperature is evaluated and displayed:

```
averageT = sumT/12
print('The average temperature is ', averageT)
```

Finally, the file is closed:

```
f.close( )
```

The complete file is:

```
# This is script Example_7_8.py
f = open( "temperatures.dat", "r")
sumT = 0
for row in f:
    data = row.split( )
    month = data[0]
    temperature = data[-1]
    word = month + " " + temperature
    sumT += float(temperature)
f.close( )
averageT = sumT/12
print('The average temperature is ' , averageT)
```

After running the script the result is 21.64

7.4.2 The Instruction readline

The instruction **readline** reads a line from a file. The format is

```
handle.readline(n)
```

The parameter n indicates how many characters in the line are read. If n is not given, the entire line is read.

Example 7.9 Reading data using readline

Consider the file `teams2.txt` from Example 7.4. This file can be read using `readline` as:

```
# This is script Example_7_9.py

# The file is opened for reading.
handle = open("teams2.txt", "r")

# The rows are read.
a = handle.readline(5)
b = handle.readline()
c = handle.readline(10)

# The data read is displayed.
print('a = ', a)
print('b = ', b)
print('c = ', c)

# The file is closed.
handle.close()
```

The result is

```
a = Brazi
b = 1 5

c = Germany 4
```

The first **readline** instruction only reads the first 5 characters in the first row. The second **readline** reads the remaining characters in the first row including the end of the line and the escape sequence "\n" which leaves an empty row. The third **readline** reads also the complete line because the third row only has seven characters. There is an empty row because of the escape sequence in the row.

7.4.3 The Instruction readlines

The instruction **readlines** reads all the lines in the file and saves them to a list.

Example 7.10 Reading data using readlines

Again the file `teams2.txt` from Example 7.4 is used. This file can be read using **readlines** as:

```
# This is script Example_7_10.py

# The file is opened for reading.
handle = open("teams2.txt", "r")

# The lines are read.
a = handle.readlines()

# The data is displayed.
print('a = ', a)

# The file is closed.
handle.close()
```

The result is a list:

> ["Brazil 5\n', 'Germany 4\n', 'Italy 4\n', 'Argentina 2\n',
> 'France 2\n', 'Uruguay 2\n', 'England 1\n', 'Spain 1']

7.5 Reading and Writing Data from and To Excel

Data created in Excel can be read in Python. A format for data files created in Excel is **csv** which stands for **comma-separated-values**. To read CSV files in Python it is necessary to import the library csv which includes the function **csv.reader** to read files and the function **csv.writer** to write files in this format.

Example 7.11 Reading data from Excel

The file **ExcelData.csv** is shown in Figure 7.1. In this Excel file, the first row is the name of the product. The following rows are the average tons bought each year. It is desired to read the data and evaluate the yearly average tons for each product. Finally, it is required to write the name and the yearly average tons for each product in another Excel file. The output file name must be **yearlyAverage.csv**.

To read the file the following steps have to be taken:

1. Import the library **csv** to be able to open the files with the extension csv. The instruction is:

   ```
   import csv
   ```

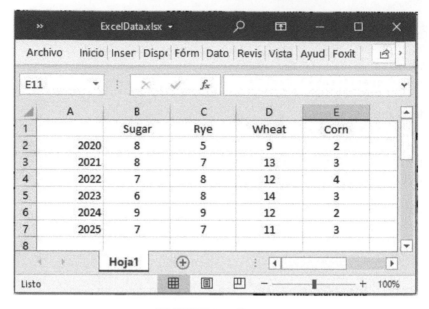

FIGURE 7.1: Input.

2. Open the file to read the information with:

```
handle = open( "ExcelData.csv", "r")
```

3. Initialize the list where each row of the data is to be stored. The list is given the name DataList:

```
DataList = [ ]
```

4. With a **for** loop the rows are read. Each row is added to the list using the instruction **append**:

```
for row in csv.reader(handle):
    DataList.append(row)
```

5. The process ends by closing the file:

```
handle.close( )
```

6. Note that the data is a list of lists; that is, a table. To display this table, the instruction `pprint` whose meaning is `pretty print`, is used. The instruction `pprint` is available in the library `pprint` and thus it has to be imported before using it. The set of instructions is:

```
import pprint
pprint.pprint(DataList)
```

The complete script is:

```
# This is script Example_7_11a.py

# Import csv and pprint.
import csv, pprint

# Open the file and initialize the list DataList:
handle = open( "ExcelData.csv", "r")
DataList = [ ]

# Read the data and place it in DataList
for row in csv.reader(handle):
    DataList.append(row)

# Close the file and print the list:
handle.close( )
pprint.pprint(DataList)
```

Now the sum of the quantities for each year is done. In the file, the first row is row No. 0, and it has the names of the products. For rows 1 to 6, the first column is the year; thus, the quantities for each product are in columns 1 to 4, and rows 1 to 6. The addition is to be written to a list that is initialized to 0. Then, the steps are:

1. Initialize the **row** with the averages to 0 and the first element of **row** as 'Averages'. This is done with:

```
row = [0.0]*len(DataList[0])
row[0] = 'Averages'
```

2. The sum of the columns is made with two nested **for** loops. The first **for** loop selects a column and the nested **for** loop adds the elements in the

column. Before the nested `for` loop the `sum` is initialized to 0. Before adding, the numbers have to be converted to floating point numbers. Recall that after read in the data from the Excel file, they are strings.

```
for c in range(1, len(row)):
    sum = 0
    for r in range(1, len(DataList)):
        sum += float(DataList[r][c])
    row[c] = sum/(len(DataList) - 1)
```

3. Create and open the file 'yearlyAverage.csv' for writing the data.

   `handle2 = open('yearlyAverage.csv', 'w')`

4. Write the elements of the first row, `DataList[0]`, which contains the names of the products. Each product name is separated by a comma. A `for` loop is used as:

```
for k in range(0, len(DataList[0])):
    handle2.write(str(DataList[0][k]))
    handle2.write(',')
```

5. A escape sequence '\n' stars a new line.

   `handle2.write('\n')`

6. The average values are stored in the list `row`. The numerical values have to be converted to strings in order to be written to the Excel file. Also a comma has to be added. A for loop is used as:

```
for k in range(len(row)):
    handle2.write(str(row[k]))
    handle2.write(',')
```

7. Finally, the file is closed:

   `handle2.close()`

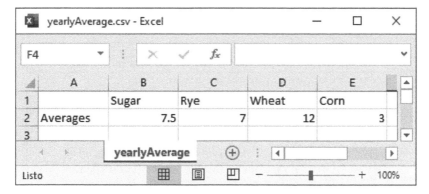

FIGURE 7.2: Output.

The complete file is:

```
# This is script Example_7_11b.py
import csv
import pprint
handle = open( "ExcelData.csv", "r")
DataList = [ ]
for row in csv.reader(handle):
   DataList.append(row)
handle.close( )
pprint.pprint(DataList)
row = [0.0]*len(DataList[0])
row[0] = 'Averages'

for c in range(1, len(row)):
     sum = 0
     for r in range(1, len(DataList)):
           sum += float(DataList[r][c])
   row[c] = sum/(len(DataList) - 1)
handle2 = open('yearlyAverage.csv', 'w')
for k in range(0, len(DataList[0])):
     handle2.write(str(DataList[0][k]))
     handle2.write(',')
handle2.write('\n')

for k in range(len(row)):
     handle2.write(str(row[k]))
     handle2.write(',')

handle2.close()
```

The results are shown in Figure 7.2.

TABLE 7.2:   Modes to handle binary files

Mode	Description
'rb'	Opens the file for read only. If the file does not exist, an error is indicated.
'wb'	Creates and opens the file to write. If the file exists, it deletes the existing data.
'ab'	Opens the file to write. If the file exists, it appends the new data at the end of the existing data.
'rb+'	Opens the file to read and write. If the file does not exist, an error message is displayed.
'wb+'	Creates and opens the file to write and read. If the file exists, the existing data is deleted.
'ab+'	Opens the file to read, write, and append. If the file does not exist, an error message is displayed.

7.6   Reading and Writing Binary Files

So far, the data used can be read by any text processor. This type of data is called ASCII data. A disadvantage of this type of data is the size of the files when handling a large amount of data. An alternative way is to store data in a binary format. This is a more efficient way to store information. Unfortunately, information stored in a binary format cannot be read by a text processor. The modes for reading, writing, and appending data in binary format are shown in Table 7.2. As before the first step is to open the file to write binary data in a new file called `testfile.txt`. This is done as:

```
handle_binary = open('testfile.txt', 'wb')
```

To write and read in a binary format, the Python distribution includes the Pickle module that has to be imported. This module has the instructions `dump` (also known as pickling) and `load` (known as unpickling) to read and write data in binary format, respectively. The syntax for the `dump` instruction is:

```
pickle.dump(data, handle)
```

An example shows how to dump data to a binary file.

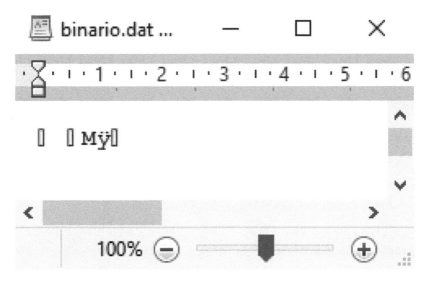

FIGURE 7.3: List A stored in binary format.

Example 7.12 Writing data to a binary file

It is desired to store the list A = [1,2,8,77, 255, 3, 0] to the file binario.dat. This can be done with the instruction dump in the following script as:

```
# This is script Example_7_12.py

# Import the necessary package.
import pickle

# Open the file for writing.
handle_binary = open('binario.dat', 'wb')

# Data to be written to the file.
A = [1, 2, 8, 77, 255, 3, 0]

# Data is witten to the file.
pickle.dump(A, handle_binary)

# The file is closed.
handle_binary.close()
```

The list A is stored in binary format as shown in Figure 7.3. This figure may change depending upon the the text editor used. However, the data stored is unchanged. In this book, the WordPad was used.

Example 7.13 Reading data from a binary file

Now the data stored in the binary file can be read with the instruction `pickle.load`. The file was created in the previous example and now the data is read with `pickle.load`. The script is:

```
# This is script Example_7_13.py

# Import the necessary package.
import pickle

# Open the file for reading.
handle_binary = open('testfile.txt', 'rb')

# Data is read from the file.
A = pickle.load(handle_binary)

# The file is closed.
handle_binary.close()

# Data read is displayed.
print("A = ", A)
```

After running this script the result is

```
A = [1, 2, 8, 77, 255, 3, 0]
```

This is the list stored in the binary file, as expected.

Example 7.14 Writing and reading binary data

Let us suppose that we want to write the following data in binary format:

$$A = \begin{bmatrix} 50 & 89 \\ 75 & 21 \end{bmatrix}$$

```
b = [54.03, 57.05, 50.8, 58.9, 89, 75.01]
```

```
B = 27
```

```
C = 'Python'
```

The data for the array could be either a list of lists or a **numpy** array. This data is entered as:

```
A = [[50, 89], [ 75, 21]]
```

```
b = [54.03, 57.05, 50.8, 58.9, 89.10, 75.01]
B = 27
C = 'Python'
```

First, the `pickle` module is imported the file `binary3.dat` is created and opened for writing binary data :

```
import pickle
id_binary = open('binary3.dat', 'wb')
```

To write the data in binary format the instruction **dump** from the library `pickle` is used. The data is an array, a list, an integer number, and a string. The data for the array could be either a list of lists or a **numpy** array.

```
pickle.dump(A, id_binary)
pickle.dump(b, id_binary)
pickle.dump(B, id_binary)
pickle.dump(C, id_binary)
```

Finally the file is closed:

```
id_binary.close()
```

The complete file to write data in binary format to file is

```
# This is script Example_7_14a.py

# Import the library pickle.
import pickle

# Open the file for writing.
id_binary = open('binary3.dat', 'wb')

# Data to be written to the file.
A = [[50, 89], [ 75, 21]]
b = [54.03, 57.05, 50.8, 58.9, 89.10, 75.01]
B = 27
C = 'Python'

# Data is written to the file.
 pickle.dump(A, id_binary)
 pickle.dump(b, id_binary)
 pickle.dump(B, id_binary)
 pickle.dump(C, id_binary)

# File is closed.
 id_binary.close()
```

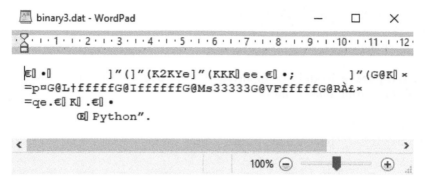

FIGURE 7.4: Binary data.

Figure 7.4 shows the data stored in binary format and open with Windows WordPad. This is because the data is written in binary format and not in ASCII code. To read the data, first the file is opened:

```
id_bin = open('binary3.dat', 'rb')
```

To read the data in binary format the `pickle.load` instruction is used. Thus, the module `pickle` has to be imported. Data must be read in the same order and with the same format that it was written. The first few lines in the file must be:

```
# Import the library pickle.
import pickle

# Read the binary data with pickle.load:
A = pickle.load(id_bin)
b = pickle.load(id_bin)
B = pickle.load(id_bin)
C = pickle.load(id_bin)
```

To display the data with `print`:
```
print("A = ", A)
print("b = ", b)
print("B = ", B)
print("C = ", C)
```

Finally, the file is closed:

```
id_bin.close()
```

The complete script to read the binary data is:

```
# This is script Example_7_14b.py

# Import the library pickle.
import pickle

# The file is opened to read:
id_bin = open('binary3.dat', 'rb')

# Read the binary data with pickle.load:
A = pickle.load(id_bin)
b = pickle.load(id_bin)
B = pickle.load(id_bin)
C = pickle.load(id_bin)

# Display the data with print:
print("A = ", A)
print("b = ", b)
print("B = ", B)
print("C = ", C)

#  The file is closed:
id_bin.close()
```

The data displayed is:

```
A = [[50, 89], [75, 21]]
b = [54.03, 57.05, 50.8, 58.9, 89.1, 75.01]
B = 27
C = Python
```

7.7   Python Instructions in Chapter 7

Table 7.3 lists the instructions introduced in Chapter 7.

7.8   Conclusions

This chapter has covered the way to store and retrieve data in a file. The data file may be created using another programming language. The different techniques are presented by using several examples. In addition, the way to retrieve and store data in Excel files is also covered.

TABLE 7.3: Python instructions in Chapter 7

Instruction	Description
close	It closes a file.
csv.reader	It reads csv files.
csv.writer	It writes csv files.
import csv	It imports the library csv.
import pprint	It imports the library pprint.
open	It opens a file for reading or writing.
pprint	It prints tables.
with	It opens files and closes them when it finishes.
write	It writes in a file.
writerow	It writes a whole row in a file.
writelines	It writes several rows to a file.

7.9 Exercises

1. Generate a list of ten integer random numbers between 1 and 100. Then, store them in a file with the name randomIntegers.dat. When finished, open the file with a word processor to check the data stored.

2. Create a file with the names of South American countries and their capital cities. The data must be previously stored in a dictionary. Then read the dictionary and store the data in a file countries.dat. Check your result in a word processor.

3. It is desired to have a file with a set of 50 randomly generated floating-point numbers. Additionally, compute the sum, the mean, and the standard deviation and store them in a dictionary. The file must be named numbers.dat.

4. The data in the previous exercise must be converted to a file numbers.csv in order to be read from Excel.

Chapter 8

Plotting in Python

8.1 Introduction

An important characteristic of any program is its ability for data visualization. This allows to see how data is behaving. For example, the behavior of the stock market can be better appreciated with a plot. Plotting trigonometric functions allows users to see that they are periodic functions. Programs written in Python can take advantage of data visualization thanks to John D. Hunter (1968-2012) who designed the module or library `matplotlib`. This module is freely available and produces very high-quality two and three-dimensional plots. Instructions for downloading `matplotlib` are available in the Appendix A. The examples presented in this chapter cover many of the typical plots used in academia and industry. For more examples, the reader can consult the `matplotlib` page: `https://matplotlib.org/stable/gallery/index.html`.

8.2 Plots in Two Dimensions

The basic instruction for plotting is `plot(x, y)` where x and y are lists of data points. When the plot function is executed, a new window is opened with the plot. The plot function is available in the sublibrary `pyplot` and must be imported prior to its use. An example shows its usage.

Example 8.1 A plot of `sin(x)`
It is desired to plot the function `sin(x)` from -2π to $+2\pi$. The list of the values of x are generated with the function `linspace` from `numpy` that has to be imported before, as:

```
import numpy as np
x = np.linspace(-2*np.pi, 2*np.pi, 100)
```

The list starts at the initial value of -2π, ends at the final value $+2\pi$ and has 100 points. The `sin(x)` function is imported from `numpy` and the y values are generated with a process called *vectorization* which consists of parallel computing methods used to perform repetitive instructions into a single vector.

DOI: 10.1201/9781003222118-8

Thus, for example, given a vector x, the sine function for each of the elements of vector x can be stored in vector y and computed as

```
y = np.sin(x)
```

The list of points for x is obtained with

```
x = linspace(-2*pi, 2*pi, 100)
```

The list for the function y is simply

```
y = np.sin(x)
```

A plot can be obtained using the package `pyplot` which is a part of `matplotlib`. Importing of `pyplot` can be done with

```
import matplotlib.pyplot as plt
```

The functions needed to obtain a plot are `plot(x, y)` where x and y are the independent and dependent lists, respectively. Additionally, to display the plot the function `show` has to be included after the `plot` instruction. Then, to obtain the plot we have to write:

```
import matplotlib.pyplot as plt
plt.plot(x, y)
plt.show()
```

The complete file is

```
''' This is script Example_8_1.py
This program plots a sin(x) function.'''

# The functions from the corresponding libraries
# are imported:
import matplotlib.pyplot as plt
import numpy as np

# The list for x is generated:
x = np.linspace(-2*np.pi, 2*np.pi, 100)

# The list for y is generated by vectorizing:
y = np.sin(x)

# Script continues next page.
```

```
# Script continued from previous page.

# The plot is created:
plt.plot(x, y)

# The plot is displayed:
plt.show()
```

The resulting plot is shown in Figure 8.1. In this plot the sine wave is called the trace. This plot is produced with 100 points for x and the number of points determines its definition. Sometimes, it may be desirable to have more points and thus obtain a higher definition plot.

The previous plot can be improved by adding a `title`, a `grid`, `ticks`, a `label`, and a `legend`.

Example 8.2 A plot of `sin(x)`
The plot from Example 8.1 is improved with a title, a grid, ticks, a label, a legend, plot limits, color change, and markers. The ticks are numbers in the axes to mark special values. The ticks and plot limits are imported from

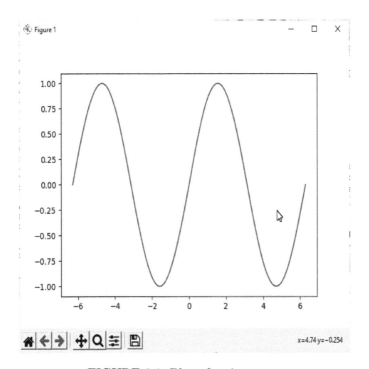

FIGURE 8.1: Plot of a sine wave.

matplotlib.pyplot. The x-axis goes from -2π to $+2\pi$ and it is desired to place ticks at -2π, $-3\pi/2$, $-\pi$, $-\pi/2$, 0, $\pi/2$, π, $3\pi/2$, and 2π. This is done with

```
# First, import the necessary functions.
from matplotlib.pyplot import xticks, yticks, xlim, ylim, grid

xticks([-2*pi, -3*pi/2, -pi, -pi/2, 0, pi/2, pi, 3*pi/2,\
       2*pi],[r'$-2\pi$', r'$-\3pi/2$', r'$-\pi$', r'$-\pi/2$',\
       r'$0$',r'$\pi/2$', r'$\pi$', r'$\3pi/2$', r'$2\pi$'])
```

The limits for the x-axis go from -7 to +7 and for the y-axis from -1.5 to +1.5 as

```
   xlim(-7, 7)
   ylim(-1.5, 1.5)
```

The title of the figure is

```
   title('Plot of a sine function')
```

The grid is added:

```
   grid()
```

For a legend located at the upper left side, it is required to have an instruction label inside the plot function as

```
   plot(x, y, label = 'sine function')
   legend(loc = 'upper left')
```

The complete file is

```
"' This is script Example_8_2.py
This program plots a sin(x) function.
It is improved with ticks, limits, title, and a legend"

# The functions from the corresponding libraries
# are imported:
from matplotlib.pyplot import plot, show
from matplotlib.pyplot import xlim, ylim, xticks, legend,
title, grid
from numpy import linspace, pi, sin

# Script continues next page.
```

```
# Script continued from previous page.

# The lists for x and y are generated with:
x = linspace(-2*pi, 2*pi, 100)
y = sin(x)

# The grid is added:
grid()

# The plot is created:
plot(x, y, label = 'sine function')
legend(loc = 'upper left')

# The x and y-axis limits are specified with:
xlim(-7, 7)
ylim(-1.5, 1.5)

# The x-ticks are added with:
xticks([-2*pi, -3*pi/2, -pi, -pi/2, 0, pi/2, pi, 3*pi/2,\
    2*pi],[r'$-2\pi$', r'$-3\pi/2$', r'$-\pi$', r'$-\pi/2$',\
    r'$0$',r'$\pi/2$', r'$\pi$', r'$3\pi/2$', r'$2\pi$'])

# The title is added:
title('Plot of a sine function')

# Finally, the plot is displayed:
show()
```

Figure 8.2 shows the resulting plot which is a better one as compared to Figure 8.1.

Example 8.3 Two functions in a plot

It is possible to get a plot with several traces corresponding to different functions. In this example the plot of the sine function, an exponential, and a parabola are plotted in the same figure. The x values are from -2π to 2π. Thus, the x vector with 100 points is given by

```
x = np.linspace(-np.pi, np.pi, 100)
```

Using vectorization, the y vectors are given by

```
y1 = sin(x)
y2 = exp(x)
y3 = x**2
```

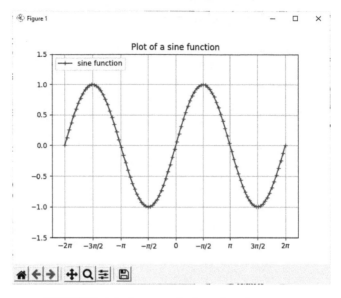

FIGURE 8.2: Improved plot of a sine wave.

These functions have to be imported before they can be used. The plotting is done with

```
plt.plot(x, y1, x, y2, x, y3)
```

now a grid is added and the plot is displayed it with

```
plt.grid()
plt.show()
```

the complete script is

```
"" This is script Example_8_3.py
This program plots a three functions in the sameplot.
It is improved with ticks, limits, title, and a legend""

# The necessary libraries are imported:
import matplotlib.pyplot as plt
import numpy as np

# Define the vector x:
x = np.linspace(-np.pi, np.pi, 100)

# Script continues at next page.
```

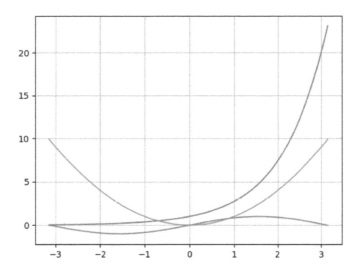

FIGURE 8.3: A plot for three functions.

```
# Script continued from previous page.

# Use vectorization to obtain the y vectors:
y1 = np.sin(x)
y2 = np.exp(x)
y3 = x**2

# The plotting is done with:
plt.plot(x, y1, x, y2, x, y3)

# A grid is added and the plot is displayed it with:
plt.grid()
plt.show()
```

The plot produced is shown in Figure 8.3. In this plot each trace is drawn in a different color, the first color corresponding to y1 is blue, the second function y2 is red, and the last one is green. A similar plot can be obtained by replacing the instruction plt.plot(x, y1, x, y2, x, y3) with three instructions plot as

```
plt.plot(x, y1)
plt.plot(x, y2)
plt.plot(x, y3)
```

TABLE 8.1:   Color palette for
the trace

Symbol	Color
b	blue
g	green
r	red
c	cyan
m	magenta
y	yellow
k	black
w	white

followed by the `plt.show()`. Every plot instruction before the `plt.show()` is plotted in the same figure.

8.2.1   Color and Marker Options

There are options to enhance a plot with color and different traces. In Figure 8.3, the colors are changed in a predetermined way as shown in Table 8.1. An example shows how to specify the color in a trace.

Example 8.4 Choosing the traces color
It is desired to change the color in a trace, for example, the traces in Figure 8.3. For the first trace it is desired to use the black color, for the second trace it is desired to use the magenta color, and for the third trace the yellow color. These changes can be done just by adding the color declaration as:

```
plt.plot(x, y1, color = 'k')
plt.plot(x, y2, color = 'm')
plt.plot(x, y3, color = 'y')
```

The script is:

```
'''This is script Example_8_4.py
This program plots a three functions in the sameplot.
It is improved with ticks, limits, title, and a legend'''

# The necessary libraries are imported:
import matplotlib.pyplot as plt
import numpy as np

# Script continues at next page.
```

```
# Script continued from previous page.

# Define the vector x:
x = np.linspace(-np.pi, np.pi, 100)
# Use vectorization to obtain the y vectors:
y1 = np.sin(x)
y2 = np.exp(x)
y3 = x**2

# The plotting is done with:
plt.plot(x, y1, color = 'k')
plt.plot(x, y2, color = 'm')
plt.plot(x, y3, color = 'y')

# A grid is added and the plot is displayed it with:
plt.grid()
plt.show()
```

The result is similar to Figure 8.3 but with different colors for the traces. Alternatively, the word `color` and the equal sign can be omitted.

TABLE 8.2:   Trace markers

Symbol	Description
-	solid line
−	discontinuous line
-.	dash and point
:	points
.	points
o	circle
v	downward pointing triangle
∧	upward pointing triangle
<	left pointing triangle
>	right pointing triangle
s	square
+	plus sign
x	cross
D	diamond
d	thin diamond
h	hexagon
p	pentagon

A trace marker and a trace style can be also added. The options are shown in Tables 8.2 and 8.3.

TABLE 8.3:   Trace style

Symbol	Description
-	solid line
–	discontinuous line
-.	dash and point
:	points

Example 8.5 Choosing the traces color and style

It is desired to change the color, the style, and the marker in a trace, for example, the traces in Figure 8.3. For the first trace, it is desired to use the red color, for the second trace it is desired to use the black color, and for the third trace the green color. The desired style is "-." for the first trace, "–" for the second one, and ":" for the last trace. For the marker, for the first trace use a circle, for the second one use a v, and for the last one use a semicolon. These changes can be done just by adding the required information as:

```
plt.plot(x, y1, 'ko-.')
plt.plot(x, y2, 'mv-')
plt.plot(x, y3, 'yo:')
```

Alternatively, the three lines can be replaced with

```
plt.plot(x, y1,'k*-.', x, y2, 'mv-', x, y3, 'yo:')
```

and the same result is obtained. The final script is:

```
"' This is script Example_8_5.py
This program plots a three functions in the same plot.
The color, style, and marker are defined here."'

# The necessary libraries are imported:
import matplotlib.pyplot as plt
import numpy as np

# Define the vector x.
# To appreciate the effect for the marker and style
# use only 10 points.
x = np.linspace(-np.pi, np.pi, 10)

# Script continues at next page.
```

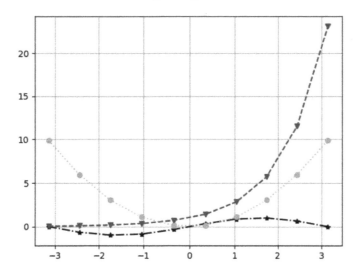

FIGURE 8.4: Plot with color, style, and markers selected.

```
# Script continued from previous page.

# Use vectorization to obtain the y vectors:
y1 = np.sin(x)
y2 = np.exp(x)
y3 = x**2

# The plotting is done with:
plt.plot(x, y1, 'ko-')
plt.plot(x, y2, 'mv-')
plt.plot(x, y3, 'yo:')

# A grid is added and the plot is displayed it with:
plt.grid()
plt.show()
```

The resulting plot is shown in Figure 8.4.

8.3   The Package seaborn

The plots implemented with matplotlib can be improved with the addition of the functions in the package seaborn. The package was designed in order to make the figure aesthetically more pleasant to look at. Some basic parameters for the plot are available in the function `set_theme` in the package `seaborn`. This function contains several parameters that make plots more appealing to see, for example, to give a color to the background, to include a grid, among other parameters. An example illustrates this.

Example 8.6 A plot using seaborn

The previous example is repeated with the `seaborn` package which has to be imported. Additionally, the function `set_theme` has to be called. These actions are done with:

```
import seaborn as sns
sns.set_theme()
```

The rest of the script remains the same, except for the `grid` function which is not needed. The final script is:

```
"' This is script Example_8_6.py
This program plots a sin(x) function.
It is improved with ticks, limits, title, and a legend"'

# The functions from the corresponding libraries are imported:
from matplotlib.pyplot import plot, show
from matplotlib.pyplot import xlim, ylim, xticks, legend    from
matplotlib.pyplot import title, grid
from numpy import linspace, pi, sin
import seaborn as sns

sns.set_theme()

# The lists for x and y are generated with:
x = linspace(-2*pi, 2*pi, 100)
y = sin(x)

# The plot is created:
plot(x, y, label = 'sine function')
legend(loc = 'upper left')

# The x and y-axis limits are specified with:
xlim(-7, 7)
ylim(-1.5, 1.5)

# Script continues at next page.
```

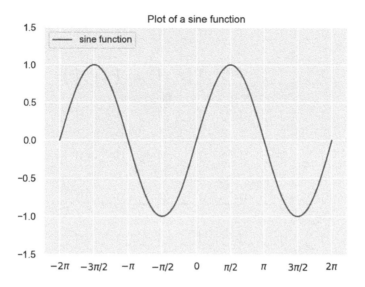

FIGURE 8.5: Plot of a sine wave with **seaborn**.

```
# Script continued from previous page.

# The x-ticks are added with:
xticks([-2*pi, -3*pi/2, -pi, -pi/2, 0, pi/2, pi, 3*pi/2,\
    2*pi],[r'$-2\pi$', r'$-3\pi/2$', r'$-\pi$', r'$-\pi/2$',\
    r'$0$',r'$\pi/2$', r'$\pi$', r'$3\pi/2$', r'$2\pi$'])

# The title is added:
title('Plot of a sine function')

# Finally, the plot is displayed:
show()
```

The resulting plot is shown in Figure 8.5

8.4 Other Two-dimensional Plots

There are many other types of two-dimensional plots such as polar, histogram, stair, stem, pie, compass, and bar plots to mention a few. This section describes these types of plots using examples.

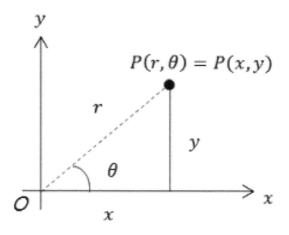

FIGURE 8.6: Polar coordinates for a point (x, y).

8.4.1   Polar Plots

In the two-dimensional polar coordinate system, each point in the plane is determined by its distance to the origin and by the angle that a line connecting the origin to the point makes with the positive x-axis. This is illustrated in Figure 8.6. The relationship between rectangular coordinates (x, y) and polar coordinates (r, θ), where $r > 0$, is

$$r = \sqrt{x^2 + y^2} \tag{8.1}$$

$$\theta = \tan^{-1}\frac{y}{x} \tag{8.2}$$

Example 8.7 A polar plot
For the function in polar coordinates

$$r = 1 + \cos(4\theta)$$

the plot for $0 < \theta < 2\pi$ can be generated with

```
''' This is script Example_8_7a.py
This program makes a polar plot.'''

# The necessary libraries are imported:
import matplotlib.pyplot as plt
import numpy as np

# Script continues at next page.
```

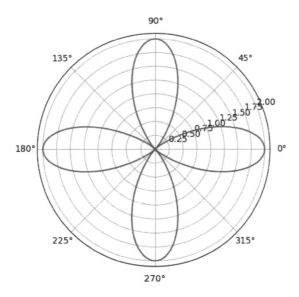

FIGURE 8.7: Polar plot for $r = 1 + \cos(4\theta)$.

```
# Script continued from previous page.

# The list for the theta values is generated with linspace.
theta = np.linspace(0, 2*np.pi, 200)

# The values for r are generated:
r = 1 + np.cos(4*theta)

# The polar plot is generated and displayed:
plt.polar(theta, r)
plt.show()
```

The polar plot is shown in Figure 8.7. There it can be seen the values for θ from 0 to 2π and the values for r from 0 to 2.00. Now it is desired to plot the polar function

$$r = \sin(4\theta)$$

that can be plotted with:

```
theta = np.linspace(0, 2*np.pi, 200)
r = np.sin(4*theta)
plt.polar(theta, r)
```

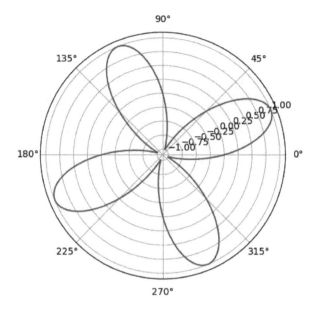

FIGURE 8.8: Polar plot for $r = \sin(4\theta)$.

The script for this polar plot is:

```
"" This is script Example_8_7b.py
This program makes a polar plot."""

# The necessary libraries are imported:
import matplotlib.pyplot as plt
import numpy as np

# The list for the theta values is generated with linspace.
theta = np.linspace(0, 2*np.pi, 200)

# The values for r are generated:
r = np.sin(4*theta)

# The polar plot is generated and displayed:
plt.polar(theta, r)
plt.show()
```

The polar plot produced is shown in Figure 8.8. There it can be seen the values for r go from -1 to $+1$. But it is known that r should be greater than 1. A negative value for r indicates that θ increases by π radians. Then, the value has to be corrected by adding π radians to the angle for those values where $r < 0$. Then, before plotting the following **for** loop is added:

```
for i in range(0, len(theta)):
    if r[i] < 0:
        r[i] = -r[i]
        theta[i] = theta[i] + np.pi
```

The final script is

```
"'' This is script Example_8_7c.py
This program makes a polar plot.'''

# The necessary libraries are imported:
import matplotlib.pyplot as plt
import numpy as np

# The list for the theta values is generated with linspace.
theta = np.linspace(0, 2*np.pi, 200)

# The values for r are generated:
r = np.sin(4*theta)

# The values are corrected if r < 0
for i in range(0, len(theta)):
    if r[i] < 0:
        r[i] = -r[i]
        theta[i] = theta[i] + np.pi

# The polar plot is generated and displayed:
plt.polar(theta, r)
plt.show()
```

The resulting plot is shown in Figure 8.9. It can be seen that the values for **r** are positive.

8.5 Pie Charts

A pie chart is used to display percentage values as **a** = [10, 25, 40, 15].
The pie chart is divided into slices. The options are labels for each slice given in a list, a shadow given in a boolean variable as **shadow** = **True**, display the value as a percentage with **autopct**='%1.2f%%' to display the percentage with two decimal places, and the start angle is given by **startangle** = **value**. A slice can be slightly separated from the pie chart with **explode** values in a list. For example, if the second slice has to be slightly exploded, then the list is

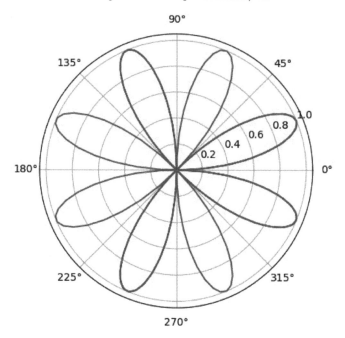

FIGURE 8.9: Polar plot for $r = \sin(4\theta)$ corrected.

explode = [0, 0.1, 0, 0.3]. Finally, the angle for the first slice separation can be specified as startangle = value in degrees

Example 8.8 A pie chart

It is required to draw a pie chart for the values in fruits = [8, 44, 16, 32]. It is also required to display the labels ['bananas', 'pears', 'apples, 'kiwis', 'mangos'], explode the second and first slices, put a shadow, display the percentage values, and start at 90°. The script is:

```
"' This is script Example_8_8.py
This program produces a pie plot.'"

# The functions from the corresponding libraries
# are imported:
import matplotlib.pyplot as plt
import numpy as np

# The list for the pie chart is:
fruits = [8, 44, 16, 32, 20]

# Script continues at next page.
```

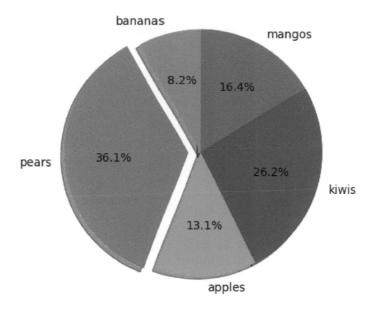

FIGURE 8.10: Pie chart for a = [8, 44, 16, 32, 20].

```
# Script continued from previous page.

# The names and options for the explode part are:
names = ['bananas', 'pears', 'apples', 'kiwis', 'mangos']
exp = [0, 0.1, 0, 0, 0]
pct = '%1.2f%%'

# The pie chart is generated and displayed:
plt.pie(fruits, labels = names, shadow = True, \
    autopct = pct, explode = exp, startangle = 90)
plt.show()
```

The pie chart produced is shown in Figure 8.10. It can be seen that the slice for bananas stars at 90°, there is a shadow, the fruit names are displayed, the percentages are displayed with two decimal digits, and the second slice corresponding to pears is exploded.

8.5.1   Histogram Plots

A histogram plot is a vertical bar graph usually used when vast amounts of statistical data have to be plotted. In this plot the number of bins may be specified and the default value is 10 bins. The instruction has to be imported from `matplotlib.pyplot` and the format is:

```
N, bins, patches = hist(data, no_bins)
```

where N is the number of data samples in each bin, `bins` is the total number of bins, and `patches` is an object that gives access to the properties of each bin.

Example 8.9 A histogram plot
It is required to plot a set of data randomly generated with a normal distribution as:

```
import numpy.random as npr
numbers = npr.normal(size = 2000)
```

The histogram can be generated with:

```
import matplotlib.pyplot as plt
N, bins, patches = plt.hist(numbers)
plt.show()
```

The complete script is:

```
"' This is script Example_8_9a.py
This program produces a histogram plot."'

# The necessary libraries are imported:
import matplotlib.pyplot as plt
import numpy.random as npr

# The random data is generated:
numbers = npr.normal(size = 100000)

# The histogram plot is generated and displayed:
N, bins, patches = plt.hist(numbers, bins = 20)
plt.show()
```

The colors of the bins may be changed and the color selected depending upon the bin size. To do this first determine the fraction of data corresponding to each bin with

```
fracs = N/N.max()
```

Next, normalize the color with

```
from matplotlib import colors
norm = colors.Normalize(fracs.min(), fracs.max())
```

Finally, assign the colors:

```
for thisfrac, thispatch in zip(fracs, patches):
        color = plt.cm.viridis(norm(thisfrac))
        thispatch.set_facecolor(color)
```

The final script is:

```
"'' This is script Example_8_9b.py
This program produces a histogram plot.
Each bin has a color assigned.'"

# The necessary libraries are imported:
import matplotlib.pyplot as plt
import numpy.random as npr
from matplotlib import colors

# The random data is generated:
numbers = npr.normal(size = 100000)

# The histogram plot is generated:
N, bins, patches = plt.hist(numbers, bins = 20)

# The fraction for each bin is calculated
# and the color is normalized:
fracs = N/N.max()
norm = colors.Normalize(fracs.min(), fracs.max())

# The color for each bin is set:
for thisfrac, thispatch in zip(fracs, patches):
        color = plt.cm.viridis(norm(thisfrac))
        thispatch.set_facecolor(color)

# The plot is displayed:
plt.show()
```

The resulting plot is shown in Figure 8.11. There it can be seen that the central bins are colored yellow and the outer bins are purple.

Another way to generate histogram plots is with **seaborn**. The function **displot** plots histograms and can smooth the histogram plot with the kernel density distribution (**kde**). The following script plots the normal distribution:

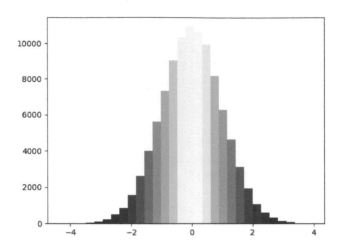

FIGURE 8.11: Histogram plot.

```
''' This is script Example_8_9c.py
This program produces a histogram plot
using seaborn.'''

# The necessary libraries are imported:
import matplotlib.pyplot as plt
import numpy.random as npr
from matplotlib import colors
import seaborn as sns
import random as rnd
from numpy.random import random

# Generate the normal distribution.
gaussian_numbers = normal(size = 1000)
x = [10 + 50*rnd.random()%(50 - 10 +1) for i in range(100)]

# Call the function with the parameters for the plot.
sns.set_theme()

# Generate the histogram with 25 bins.
sns.displot(gaussian_numbers, bins = 25)

# Display the histogram.
plt.show()
```

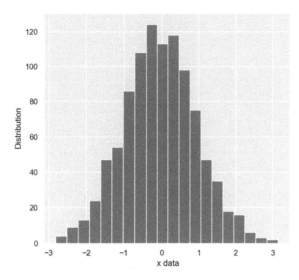

FIGURE 8.12: Histogram plot created with `seaborn`.

The resulting plot is shown in Figure 8.12. If a histogram smoothing is desired the `kde` parameter and its value have to be added. This is done by replacing the `displot` line with

```
sns.displot(gaussian_numbers, bins = 25, kde = True)
```

The new plot with smoothing is shown in Figure 8.13.

8.5.2 Stem Plots

A stem plot draws a vertical line from the horizontal axis and places a marker on top of it. The function plotted is stored in a list. The default marker is a circle but it can be changed in `markerfmt` by any marker from Table 8.2. The vertical line style can be changed with `linefmt` for any style line in Figure 8.3.

Example 8.10 A stem plot
It is required to plot the sine function in a stem plot. The range for `x` is from 0 to 2π. The stem plot must have 30 points. The `stem` instruction is used and it has to be imported from `matplotlib.pyplot`. First, the libraries required are imported:

```
from matplotlib.pyplot import show, stem
from numpy import linspace, pi, sin
```

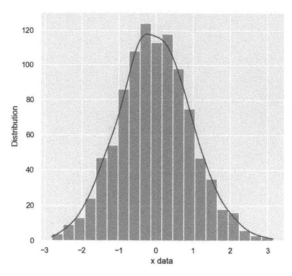

FIGURE 8.13: Histogram plot with **seaborn** with smoothing.

The x-values are generated with **linspace** and then the y-values are generated by vectorizing as:

```
x = linspace(-2*pi, 2*pi, 30)
y = sin(x)
```

The stem plot is generated with **linefmt** = ':' and **markerfmt** = 'x'. Finally, the plot is displayed with A **show()**:

```
stem(x, y, linefmt = ':', markerfmt = 'x')
show()
```

The complete file is:

```
''' This is script Example_8_10.py
This program produces a stem plot.'''

# The necessary libraries and functions are imported:
from matplotlib.pyplot import show, stem
from numpy import linspace, pi, sin

# The x and y-values are generated:
x = linspace(-2*pi, 2*pi, 30)
y = sin(x)

# Script continues at next page.
```

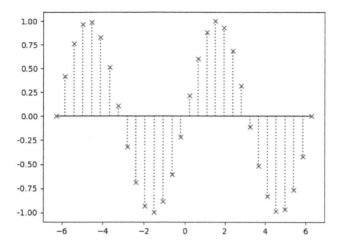

FIGURE 8.14: Stem plot.

```
# Script continued from previous page.
# The stem plot is generated and displayed:
stem(x, y, linefmt = ':', markerfmt = 'x')
show()
```

The resulting plot is shown in Figure 8.14. There it can be seen that the vertical lines are dotted and the marker is a cross.

8.5.3 Scatter Plots

A scatter plot is similar to the stem plot but there are no lines connecting the data points with the horizontal axis. A scatter plot uses the keyword `scatter` available from `matplotlib.pyplot`. The format is

```
scatter(x, y, s, c, alpha, marker, parameters)
```

where `x`, `y` are the coordinates, `s` is the size, `c` specify the color, `marker` is the shape, `alpha` with a value between either 0 for a transparent or 1 for an opaque marker, and parameters specify other parameters for the data points. The parameters just mentioned are given for each point in the plot and thus, they can be `numpy` arrays. An example is now shown.

Example 8.11 Scatter plot
Consider the set of points randomly generated by

```
N = 20
x = np.random.rand(N)
y = np.random.rand(N)
```

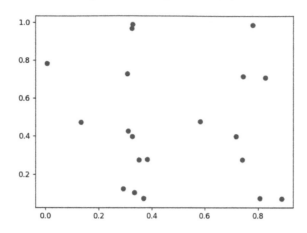

FIGURE 8.15: Scatter plot.

the scatter plot is generated with

```
plt.scatter(x, y, marker = 'x')
plt.show()
```

The complete script is

```
"'' This is script Example_8_11.py
This program makes a scatter plot.'"

# The necessary libraries are imported:
import matplotlib.pyplot as plt
import numpy as np

N = 20
x = np.random.rand(N)
y = np.random.rand(N)

# The scatter plot is generated with:
plt.scatter(x, y, marker = 'x')
plt.show()
```

The resulting plot is shown in Figure 8.15. Now, more parameters can be included; for example, if the line for the scatter plot is replaced by:

```
colors = np.arange(N)
alpha = 0.7
s1 = abs(np.random.rand(N))
plt.scatter(x, y, s = s1, c = colors, alpha = 0.7, marker = 'x')
```

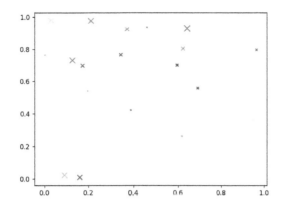

FIGURE 8.16: Scatter plot with parameters.

The result is shown in Figure 8.16 where the different colors, opacity, and size of the markers can be noted.

8.6 Multiple Figures

In this section it is shown how to create multiple figures. There are two possibilities, namely, each plot in a separate figure and several plots in a single figure.

8.6.1 Plots in Separate Figures

In each of the figure plots generated so far, the window name is **Figure 1**. This is the default figure number. If another plot is generated in the same program, the previous plot is deleted and a new **Figure 1** is opened. In order to have several figure windows open it is necessary to create new figure windows with the instruction **figure(number)** that generates a new figure window with the name **Figure window**. Each time **figure()** is used a new figure window with consecutive numbering is displayed. If some number is written inside the parenthesis, then a figure window with that number is created.

Example 8.12 Two plots in two separate windows
It is required to plot the sine and cosine functions in two separate figure windows. The figure window numbers are **Figure 5** and **Figure 10**. This is done first by importing the necessary libraries and functions with:

```
from matplotlib.pyplot import plot, show, figure
from numpy import linspace, pi, sin, cos
```

Then, generating the x-values with `linspace` and the functions `sin(x)` and `cos(x)` are evaluated by vectorizing:

```
x = linspace( -pi, pi, 256)
y = sin(x)
z = cos(x)
```

Next, the functions are displayed in figure windows 5 and 10 as follows:

```
figure(5)
plot(x, y, label = 'sine')
legend(loc = 'upper left')
figure(10)
plot(x, z, label = 'cosine')
legend(loc = 'upper right')
```

Finally, the figures are displayed:

```
show()
```

The complete script is:

```
''' This is script Example_8_12.py
This program produces two figure windows.'''

# The necessary libraries and functions are imported:
from matplotlib.pyplot import plot, show, legend, figure
from numpy import linspace, pi, sin, cos

# The x, y, and z-values are generated:
x = linspace( -pi, pi, 256)
y = sin(x)
z = cos(x)

# The two figure windows are generated
figure(5)
plot(x, y, label = 'sine')
legend(loc = 'upper left')

# Script continues at next page.
```

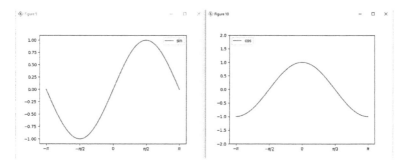

FIGURE 8.17: Two windows with figures.

```
# Script continued from previous page.

figure(10)
plot(x, z, label = 'cosine')
legend(loc = 'upper right')

# The figures are displayed:
show()
```

The resulting plot is shown in Figure 8.17. It can be seen that there are two windows with names **Figure 5** and **Figure 10**.

8.6.2 Subplots

A difference from the previous way to plot two or more figures is the use of subplots. In this case, two or more plots are fitted in a single figure window and the instruction `subplot`, imported from `matplotlib.pyplot`, is used. The subfigures are placed using an array style, for example, if there are four plots they can be placed in two rows, two plots by row. Then, it is said that the arrangement is 2 by 2 and a third number gives the position of the subplot in the arrangement. For the `subplot(2, 2, 3)`, the plot is placed in the second row on the left. The `subplot(2, 2, 2)` places the plot in the first row to the right which corresponds to the second position. The format for the instruction `subplot` is:

    subplot(m, n, k)

where m and n are the number of rows and columns, respectively, and k is the position in the array from left to right and from top to bottom.

- m is the number of rows.
- n is the number of columns.
- k is the kth subplot in consecutive numbering or position in the plot.

Example 8.13 Use of subplots
Four plots can be placed in two rows. The functions to plot are the $y1 = \sin(x)$ and $y2 = \cos(x)$ from the previous example, and the functions $y3 = x^2$ and $y4 = x^3$. The libraries and functions to be imported are

```
from numpy import linspace, pi, sin, cos
from matplotlib.pyplot import plot, show, grid
from matplotlib.pyplot import legend, subplot
```

The x-values are computed and the functions are evaluated:

```
x = linspace( -pi, pi, 256)
y1 = sin(x)
y2 = cos(x)
y3 = x**2
y4 = x**3
```

The subplots are indicated and the functions are plotted with a legend and a grid:

```
subplot(2, 2, 1) # k = 1 First row left side.
plot(x, y1, label = 'sin')
legend(loc = 'upper left')
grid()

subplot(2, 2, 2) # k = 2 First row right side.
plot(x, y2, label = 'cos')
legend(loc = 'upper left')
grid()

subplot(2, 2, 3) # k = 3 Second row left side.
plot(x, y3, label = 'x∧2')
legend(loc = 'upper center')
grid()

subplot(2, 2, 4) # k = 4 Second row right side.
plot(x, y4, label = 'x∧3')
legend(loc = 'upper left')
grid()
```

Finally, the plots are displayed:

```
show()
```

The complete script is

```
"' This is script Example_8_13.py
This script produces four subplots."'

# The necessary libraries and functions are imported:
from numpy import linspace, pi, sin, cos
from matplotlib.pyplot import plot, show, grid
from matplotlib.pyplot import legend, subplot

# The x-values, y1, y2, y3, and y4 are calculated:
x = linspace( -pi, pi, 256)
y1 = sin(x)
y2 = cos(x)
y3 = x**2
y4 = x**3

# The subplots are generated with label, legend, and grid:
subplot(2, 2, 1) # k = 1 First row left side.
plot(x, y1, label = 'sin')
legend(loc = 'upper left')
grid()

subplot(2, 2, 2) # k = 2 First row right side.
plot(x, y2, label = 'cos')
legend(loc = 'upper left')
grid()

subplot(2, 2, 3) # k = 3 Second row left side.
plot(x, y3, label = 'x∧2')
legend(loc = 'upper center')
grid()

subplot(2, 2, 4) # k = 4 Second row right side.
plot(x, y4, label = 'x∧3')
legend(loc = 'upper left')
grid()

# The figures are displayed:
show()
```

A run of this script produces the plots shown in Figure 8.18.

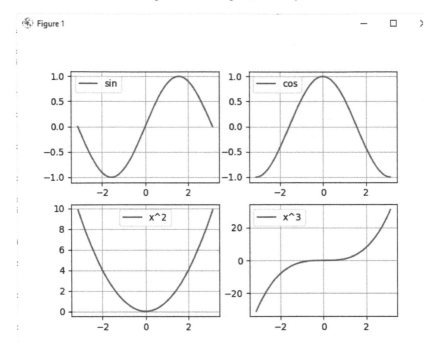

FIGURE 8.18: Four plots in a figure.

8.7 Three-Dimensional Plots

Three-dimensional plots require taking values in two variables that are generally x and y to give values to a z-axis. The three-dimensional plot can take on several forms and options. Several examples show these different types of plots. There are other plots that are considered in this section although they are not three-dimensional but form part of them, such as the quiver and the contour plot.

Example 8.14 Plot of a parametric curve
The plot of a parametric curve requires to write the equations in terms of a parameter. The parametric plot is formed with the `plot` instruction. The fact that a 3D plot is required is given by the `axes` instruction as

```
axes(projection = '3d')
```

This instruction is imported from `matplotlib.pyplot`. For example, to plot the equations:

$$z = r$$
$$x = r * sin(\theta)$$
$$y = r * cos(\theta)$$

where the variable `r` takes values given by a `linspace` instruction. In addition, a legend, and axes labels are added. The parametric plot is obtained with the following file:

```
''' This is script Example_8_14.py
This program makes a 3D plot.'''

# Import the required libraries and functions:
from matplotlib.pyplot import title, show, plot, legend, axes
from numpy import linspace, sin, cos, pi

# Generate the 3D axes:
axes(projection = '3d')

# Define the parameter theta and the x, y, z values:
theta = linspace(-4*pi, 4*pi, 100)
r = linspace(0, pi, 100)
z = r
x = r*sin(theta)
y = r*cos(theta)

# Plot and display the graph:
plot(x, y, z, label = 'Parametric plot')
legend( )
show( )
```

The resulting three-dimensional plot is shown in Figure 8.19.

Example 8.15 Surface plot
A surface plot requires the calculation of a mesh with `meshgrid` where the range must be specified with

```
X = arange(-5, 5, 0.25)
Y = arange(-5, 5, 0.25)
X, Y = meshgrid(X, Y)
```

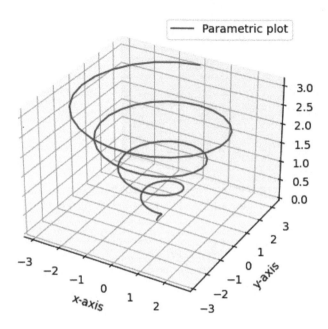

FIGURE 8.19: Three-dimensional parametric plot.

The **arange** and **meshgrid** instructions must be imported from **numpy**. The **meshgrid** instruction creates the values for the points X, Y where the desired function is going to be evaluated. For this example, the function to be plotted is

$$f(\text{ X, Y}) = \sin(\text{X}^2 + \text{Y}^2)$$

The Python instruction is

```
Z = np.sin(np.sqrt(X**2 + Y**2))
```

A surface plot is produced with

```
ax.plot_surface(X, Y, Z)
```

The plot can be colored with a color map using the instruction **cmap** inside the **plot_surface** instruction. A list of the options available for the color map is available at https://matplotlib.org/stable/tutorials/colors/colormaps.html. For this example, the color map 'gist_rainbow' is used. In addition, to improve the image sampling the parameter **antialiased** is set to **False**. The instruction **plot_surface** is then:

```
ax.plot_surface(X, Y, Z, cmap = 'gist_rainbow', antialiased=False)
```

There are two parameters that improve the final plot. They are the step size in each direction. The parameters are `rstride` and `cstride`. The default value is 10, but an improvement in the plot can be achieved by setting them to a lower value, for example, setting both of them to unity. Then the `plot_surface` instruction is

```
surf = ax.plot_surface(X, Y, Z, cmap = 'gist_rainbow', \
       rstride = 1, cstride = 1, antialiased = False)
```

The complete script is:

```
"' This is script Example_8_15.py
This program makes a 3D surface plot.'"

# Import the required libraries and functions:
import matplotlib.pyplot as plt
import numpy as np

# Create the 3D axes
ax = plt.axes(projection = '3d')

# Generate the X, Y, Z values
X = np.arange(-5, 5, 0.1)
Y = np.arange(-5, 5, 0.1)
X, Y = np.meshgrid(X, Y)
Z = np.sin(np.sqrt(X**2 + Y**2))

# The surface plot is generated and displayed
surf = ax.plot_surface(X, Y, Z, rstride = 1,\
    cmap = 'gist_rainbow', antialiased = False)
plt.show()
```

The resulting three-dimensional plot is shown in Figure 8.20. The plot can have added a color bar before displaying the plot, the instruction to plot the color bar is

```
plt.colorbar(surf, shrink = 0.5, aspect = 5)
```

The surface plot now with the color bar added is shown in Figure 8.21.

The surface plot can be observed from different points of view by simply placing the pointer over the plot, pressing the left button and dragging the

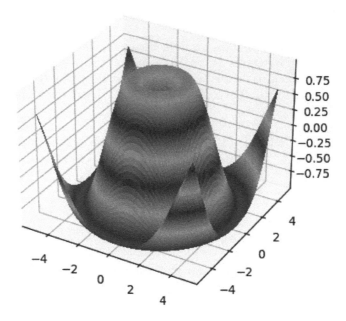

FIGURE 8.20: Three-dimensional surface plot.

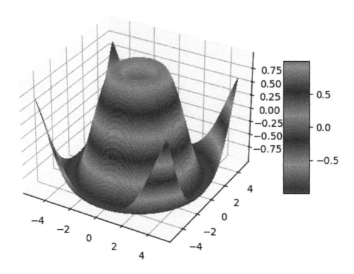

FIGURE 8.21: Three-dimensional surface plot with color bar.

pointer. This allows the observation of the plot from where it is desired to be seen.

Example 8.16 Wireframe Surface Plot

A wireframe surface plot requires the same steps as the previous example but changes the word surface to wireframe. To better appreciate the result, the instruction arange's step is changed from 0.25 to 0.5.

The complete script is:

```
"' This is script Example_8_16.py
This program makes a 3D wireframe plot."'

# Import the required libraries and functions:
import matplotlib.pyplot as plt
import numpy as np

# Create the 3D axes
ax = plt.axes(projection = '3d')

# Generate the X, Y, Z values
X - np.arange(-5, 5, 0.5)
Y = np.arange(-5, 5, 0.5)
X, Y = np.meshgrid(X, Y)
Z = np.sin(np.sqrt(X**2 + Y**2))

# The wireframe plot is generated and displayed
wire = ax.plot_wireframe(X, Y, Z, antialiased = False)
plt.show()
```

The wireframe plot is shown in Figure 8.22.

Example 8.17 Quiver plot
A quiver plot is a vector field plot. It is a two-dimensional plot related to a three-dimensional plot. It has two pairs of components. The first pair [x, y] is the position of each arrow vector and the second pair u, v is the direction. The format is

```
quiver([x, y], u, v)
```

As an example, the magnetic field due to a current in a wire is plotted. The magnetic field can be plotted using the equations of the components of the

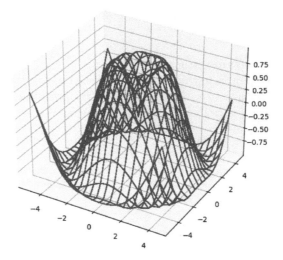

FIGURE 8.22: Wireframe plot.

normalized magnetic field vector. They are plotted on a two-dimensional plot to show how the field behaves around the wire:

$$B_x = \frac{-y}{x^2 + y^2}$$
$$B_y = \frac{x}{x^2 + y^2}$$

In the following file, these equations are evaluated and the field is as shown in Figure 8.23.

```
''' This is script Example_8_17.py
This program plots the magnetic field.'''

from pylab import *
xmax = 15.0
xmin= -xmax
NX = 7
ymax = 15.0
ymin = -ymax
NY = 7

# Script continues at next page.
```

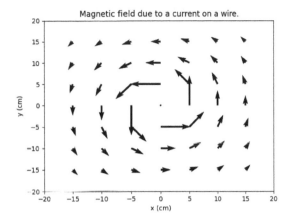

FIGURE 8.23: Quiver plot for the magnetic field around a wire. The current goes out of the page.

```
# Script continued from previous page.

# Vector components and points
x = linspace(xmin, xmax, NX)
y = linspace(ymin, ymax, NY)
X, Y = meshgrid(x, y)
S2 = X**2 + Y**2 # This is the radius squared.

# Magnetic field. The value 0.001 is to avoid division by 0.
Bx = -Y/(S2 + 0.001)
By = +X/(S2 + 0.001)

# Plot the vector field:
figure( )
QP = quiver(X, Y, Bx, By)

# Axes limits
dx = (xmax - xmin)/(NX - 1.)
dy = (ymax - ymin)/(NY - 1.)
axis([xmin - dx, xmax + dx, ymin - dy, ymax + dy])
title('Magnetic field due to a current on a wire.')
xlabel('x (cm)')
ylabel('y (cm)')

# Display the plot:
show( )
```

The magnetic field is shown in Figure 8.23. It can be seen that the magnetic field decreases in magnitude as the distance from the wire, which is located in the center of the plot, increases. Also, since the current through the wire is pointing outwards, the vector field follows the right-hand rule.

Example 8.18 Contour plot
Sometimes it is desired to make a two-dimensional projection on the x, y plane of a three-dimensional surface. This is done by plotting the contour which can be done with the instruction contour. The format is:

contour(X, Y, Z3, No. of ContourLines, cmap = 'choose color')

For the function

$$Z = \sin(X/3) + \cos(Y/4)$$

for values of X, Y in the range -3π and 3π, a surface plot can be obtained with

```
''' This is script Example_8_18a.py
This program makes a contour plot.'''

import numpy as np
# import matplotlib.ticker as ticker
import matplotlib.pyplot as plt
# Define the surface:
delta = 0.25
edge1 = np.pi*6
x = np.arange(-edge1, edge1, delta)
y = np.arange(-edge1, edge1, delta)
X, Y = np.meshgrid(x, y)
Z = np.sin(X/3) + np.cos(Y/4)

# Basic surface plot
ax = plt.axes(projection = '3d')
ax.plot_surface(X, Y, Z, cmap = 'binary')
plt.show()
```

The surface plot is shown in Figure 8.24. The surface was plotted with a color map 'binary' where the lowest values are very clear and the highest values are darker. Now, a contour plot for this surface plot is obtained by replacing the instructions

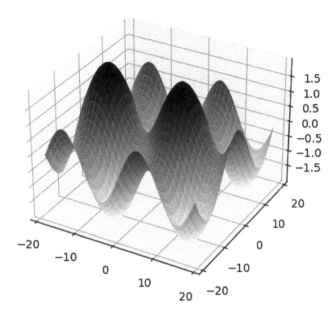

FIGURE 8.24: A surface plot

```
# Basic surface plot
ax = plt.axes(projection = '3d')
ax.plot_surface(X, Y, Z3, cmap = 'binary')
```

with the following instructions

```
# Basic contour plot
CS = plt.contour(X, Y, Z3, 15, cmap = 'binary')
plt.clabel(CS, fontsize = 10)
```

The script to obtain the contour plot is

```
"' This is script Example_8_18b.py
This program makes a contour plot.'"

import numpy as np
import matplotlib.pyplot as plt

# Script continues at next page.
```

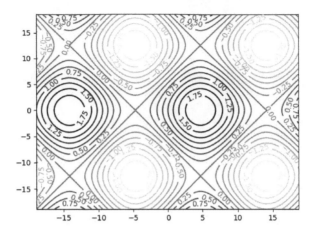

FIGURE 8.25: A contour plot

```
# Script continued from previous page.

# Define the surface
delta = 0.25
edge1 = np.pi*6
x = np.arange(-edge1, edge1, delta)
y = np.arange(-edge1, edge1, delta)
X, Y = np.meshgrid(x, y)
Z = np.sin(X/3) + np.cos(Y/4)

# Basic contour plot
# The number of contours is 15.
CS = plt.contour(X, Y, Z, 15, cmap = 'binary')
# clabel adds values to the contour lines.
plt.clabel(CS, fontsize = 10)
plt.show()
```

The contour plot is shown in Figure 8.25. The light gray lines correspond to lower values of the function and the darker levels correspond to high values. The `clabel` instruction adds values to the contours.

    The next plot adds a contour plot to the surface plot. The instructions that have to be added to the surface plot script are (after the surface plot instruction and before the `show()` instruction):

```
ax.contour(X, Y, Z, zdir = 'z', cmap = 'binary', offset = -2)
plt.clabel(CS, inline = True, fontsize = 10)
```

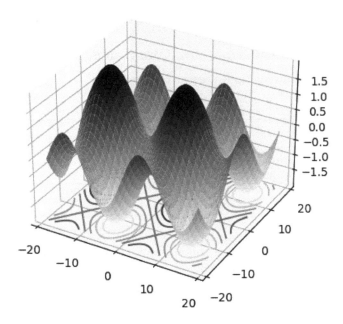

FIGURE 8.26: A surface plot with a contour plot.

The final script is

```
"' This is script Example_8_18c.py
This program makes a surface plot with a contour."'

import numpy as np
import matplotlib.pyplot as plt
# Define the surface
delta = 0.25
edge1 = np.pi*6
x = np.arange(-edge1, edge1, delta)
y = np.arange(-edge1, edge1, delta)
X, Y = np.meshgrid(x, y)
Z = np.sin(X/3) + np.cos(Y/4)

# Basic surface plot
ax = plt.axes(projection = '3d')
CS = ax.plot_surface(X, Y, Z, cmap = 'binary')
ax.contour(X, Y, Z, zdir = 'z', cmap = 'binary', offset = -2)
plt.clabel(CS, inline = True, fontsize = 10)
plt.show()
```

The surface plot with a contour is shown in Figure 8.26. In this case the contour plot does not display values for each contour.

8.8   Python Instructions for Chapter 8

Table 8.4 gives the instructions used in Chapter 8.

TABLE 8.4: Python instructions for Chapter 8

Instruction	Description
arange	Creates a list according to the parameters.
aspect	Plot's aspect ratio.
autopct	Value spacing in a polar plot.
Axes3D	Axis for a 3D plot.
cmap	Indicates the color map.
color	Color for the plot trace.
colorbar	Adds a color bar.
contour	Implements a contour plot.
coolwarm	Color of the plot.
cstride	Parameter of the 3D.
displot	Generates a histogram plot with **seaborn**.
figsize	Figure size.
figure	Creates a new figure window.
grid	Generates a grid for the plot.
hist	Generates a histogram plot.
label	Label for the plot.
legend	Adds a legend to the figure.
linewidth	Changes the thickness of the trace.
linspace	Generates values for the independent variable.
scatter	Generates a points plot.
coolwarm	A library to improve plots.
stem	Generates a points plot with a connection to the x-axis.
ß loc	Position of **legend** in the figure.
matplotlib	Library for plotting functions.
meshgrid	Grid for 3D plots.
mpl_toolkits	Library for 3D.
normal	Normal statistical distribution.
numpy	Numeric library.
pie	Generates the pie plot.
plot	Basic instruction for plotting.
plot_surface	Basic instruction for surface plotting.
pyplot	Library of plotting functions.
polar	Generates polar plot.
projection	Generates the projection for a 3D plot.

(Continues next page.)

TABLE 8.4: Python instructions in Chapter 8 (*Continued.*)

Instruction	Description
quiver	Generates a quiver plot.
rstride	Parameter for a 3D plot.
show	Displays the plot.
size	Size of the plot.
stem	Plot of points.
subplot	Generates several plots in the same figure.
title	Title of the plot.
upper left	Position of the legend.
upper right	Position of the legend.
wireframe	Mesh plot.
xlim	Limit of the x-coordinate.
xticks	Values of the x-coordinate..
ylim	Limit of the y-coordinate.
zdir	Direction of z for visualization.
zlim	Limit of the z-coordinate.

8.9 Conclusions

This chapter has presented the different ways to visualize data in Python. This is one of the most important characteristics of this programming language. Different types of plots were described and examples were given. The libraries used were `matplotlib`, `pyplot`, `seaborn`, and `azw`. For more information on the `pyplot` parameters please go the the pyplot web site [1].

8.10 Exercises

1. Plot the points y = [2, -2, 3, 8, 9, 0, 1] using the method `plot`.

2. Plot the points given by the dictionary {1:1, 2:1, 3:5, 4:9, 5:-10}. The keys are the x-coordinates and the values are the y-coordinates.

3. Plot the trigonometric function $\cos(x)$ from -2π a $+2\pi$. Use `linspace` with 100 points.

4. Repeat the previous exercise using only 10 points. Compare the results.

5. Plot the function $f(x) = x^2$ from $x = $ -3 to $x = $ +3. Add appropriate legend and label instructions. `legend` and an appropriate `label`.

6. Plot the function $\log(x)$ for $x > 0$. Use `linspace` for the x-values. Choose an appropriate range for the x-values.

7. Repeat the previous exercise for $\log10(x)$.

8. Repeat the two previous exercises using `logspace` instead of `linspace`. Compare the results.

9. Plot the Chebyshev polynomial $C_2(x) = x^3 - 2x$ in the range -2 to +2.

10. In the same figure obtain the plots for $f(x) = x^2 + 1$, $g(x) = \sqrt{x}$, and $h(x) = \operatorname{sen}^2(x)$.

11. Obtain the polar plot for the function

$$r = \frac{2}{1 - r\cos\theta}$$

12. Obtain the pie plot for the list `[23, 45, 17, 15]`.

13. Obtain the stem plot for the list `[0.3, 4.2, -1, 0, 7.8 ]`.

14. Obtain a 3D parametric plot for the equations

$$z = 1$$
$$x = r * sin(\theta)$$
$$y = r * cos(\theta)$$

15. Obtain a surface plot for the function

$$z = \frac{\sin\sqrt{x^2 + y^2}}{x^2 + y^2}$$

in the range for `(x, y)` from -10 to 10.

16. Obtain a `wireframe` plot $\sin(\sqrt{|xy|})$.

17. To the `wireframe` plot from the previous exercise add a lateral color bar.

18. Obtain a `contour` plot for

$$f(x, y) = \frac{1}{x^2 + y^2 + 1}$$

19. Obtain a `quiver` plot for the electric field with the coordinates:

$$\mathbf{E} = \left(\frac{\cos\theta}{\sqrt{x^2 + y^2}}, \frac{\sin\theta}{\sqrt{x^2 + y^2}} \right)$$

8.11   Selected Bibliography

1. https://matplotlib.org/stable/tutorials/introductory/pyplot.html#sphx-glr-tutorials-introductory-pyplot-py

Chapter 9

Optimization

In this chapter, it is shown how optimization problems can be solved using Python and an additional package called `SciPy`, short for *Scientific Python*. The available functions, some of which are used in examples in this chapter, allow the solving of linear and nonlinear problems with and without restrictions, using objective functions.

9.1   Introduction

In solving problems that involve mathematical equations, it is very common for the problem to have either too many variables, implicitly defined variables, or an equation that describes the problem. These are some of the few cases where the use of optimization techniques is required. These techniques are based on search algorithms which allow reaching one or several solutions to a problem that is analytically impossible or very difficult to solve. The application of optimization techniques is very broad and covers problems in all areas of science. Python can incorporate a package that includes functions for performing optimization for a large number of functions. The user has the task of finding the most appropriate way to solve a given problem. This chapter is a brief introduction to some of the functions for optimization in `SciPy`, so that users can apply them in solving their problems. Several examples illustrate how to solve these problems using Python and `SciPy`. `SciPy` is a package that must be installed using pip as described in the Appendix.

9.2   Optimization Concepts

Optimization refers to the best way to solve a problem having a set of restrictions and conditions. Problems, restrictions, and conditions are defined by mathematical equations. The following example (9.1) deals with how to introduce variables, functions, and parameters that are required in an optimization process.

DOI: 10.1201/9781003222118-9

Example 9.1 Optimizing the volume of a box

A typical optimization problem presented in Differential Calculus courses after studying the subject of maxima and minima is how to obtain a box with an open top. This box should be formed from a rectangular piece of cardboard whose sides are a and b. If a square at each of the corners is cut and the sides folded up, the box is formed. The aim is that the volume of the box is maximized. In this problem, the squares that are at the corners will have a side x, in such a way that the box will have a height x and sides $A\text{-}2x$ and $B - 2x$, as shown in Figure 9.1. The volume is given by

$$V = (A - 2x)(B - 2x) = 4x^3 - 2(A + B)x^2 + ABx$$

and the area of the rectangular piece of cardboard must be

$$\text{Area} = AB$$

The volume is called the target function or goal function which in this case is to be maximized. The area that remains constant is called the restriction function, which in this example is an equality constraint. The variable x is the design variable.

9.2.1 Parameters, Variables, and Functions

Performing an optimization by any of the traditional methods requires the formalization of the mathematical description of the problem. Below is a description of the variables and parameters involved in an optimization process.

- **Design variables**. With them we define a design in particular. In searching for a particular solution, variables can change in value within a prescribed range. The set of variables is known as a vector design and is denoted by $x = [x_1, x_2, ..., x_n]$.

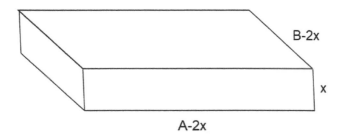

FIGURE 9.1: Volume of a box.

- **Design parameters**. These are those constant amounts which are compared with each design and that do not change during the optimization process. They are denoted with $P = [p_1, p_2, ..., p_n]$.

- **Design functions**. These are functions that contain design information. They are calculated with the values from the design variables and the design parameters. These functions can represent design objectives or restrictions.

- **Objective functions or goal functions**. These are the functions that define the objective or goal of the problem. It can be a problem about minimizing or maximizing a function. In this chapter, we will use the case of a single objective function or goal function.

- **Restriction functions**. These functions depend on the design variables. Usually, there are several restriction functions. These functions are compared with the design requirements.

- **Inequality constraints**. They are the constraints that must satisfy the problem. They are more difficult to satisfy than the equality constraints. They are denoted by the matrix equation $Ax \leq b$.

- **Equality constraints**. They are similar to the inequality constraints but are easier to fulfill. They are denoted by the matrix equation $A_{eq}x = b_{eq}$.

- **Lateral constraints**. They are used to specify the range of the design variables. They correspond to the lower and upper bounds of the design variables x.

- **Initial conditions**. They are the values of the design variables that are at the beginning of the optimization.

9.3 General Format of the Optimization Process

The definitions of parameters, variables, and functions allow the setup of the design process. This will be made up of the following components:

- Minimize objective or goal function: $f(x_1, x_2, ..., x_n)$

- Initial conditions: $x_{o1}, x_{o2}, ..., x_{on}$

Subject to

$$h_1(x_1, x_2, \ldots, x_n) = 0$$
$$h_2(x_1, x_2, \ldots, x_n) = 0$$
$$\vdots$$
$$h_m(x_1, x_2, \ldots, x_n) = 0$$

- Equality constraints: $A_{eq} = b_{eq}x$

Subject to

$$H_1(x_1, x_2, \ldots, x_n) < 0$$
$$H_2(x_1, x_2, \ldots, x_n) < 0$$
$$\vdots$$
$$H_k(x_1, x_2, \ldots, x_n) < 0$$

- Inequality constraints: $A_{ineq} \leq b_{ineq}x$

Bounded by the lower and upper bounds of the variables.
Subject to nonlinear equality and inequality conditions.

The user's task is to set the optimization problem in terms of this format.

9.4   Optimization with Python

Python developers have created many libraries for optimization. One of such packages is `SciPy` that includes, among many other sets of functions, a set of functions to implement different types of optimization strategies that can be used to solve problems in continuous or discrete time. The package includes the following functions (this list is not exhaustive):

- unconstrained minimization multivariate (`minimize`),
- unconstrained minimization univariate (`minimize_scalar`),

- bounded minimization (`bounded`),

- linear programming (`linprog`),

- Nelder-Mead (`nelder-mead`) simplex algorithm,

- Broyden-Fletcher-Goldfarb-Shanno (`BFGS`) algorithm,

- Newton-Conjugate-Gradient (`Newton-CG`) algorithm,

- Trust-Region Newton-CG (`trust-ncg`) algorithm,

- Trust-Region Truncated Generalized Lanczos/CG (`trust-krylov`) algorithm,

- Trust-Region Nearly Exact (`trust-krylov`) algorithm,

- Sequential Least SQuares Programming (`SLSQP`) algorithm,

- least-squares (`least-squares`) minimization.

They can be used to find optimal solutions, perform equilibrium analysis, balance multiple design alternatives, and incorporate methods for the optimization of algorithms and models. The methods used are within the following techniques:

- Tools for defining and solving optimization problems as well as monitoring the progress towards the solution.

- Functions for nonlinear and multi-objective optimization.

- Functions for performing optimization using nonlinear least squares, data fitting and nonlinear equations techniques.

- Methods for solving linear and quadratic programming problems.

- Methods for solving binary integer programming problems.

- Parallel computing support in some methods with nonlinear constraints.

By means of examples, the chapter shows how to use some of these functions to solve optimization problems.

9.5 The `minimize` Function

The first technique shown for minimization is the function `minimize`. This function has the following format:

```
minimize(fun, x0, args, method, jac, hess, hessp, bounds,
    constraints = (), tol, callback, options)
```

where the parameters may be:

- **fun** (callable) objective function to be minimized,

- **x0** (ndarray) initial guess,

- **args** (tuple, optional) extra arguments of the objective function and its derivatives (jac, hes),

- **method** (str, optional) optimization methods,

- **jac** (bool or callable, optional) Jacobian (gradient),

- **hess, hessp** (callable,optional) Hessian (2nd-order grad.) and Hessian-vector product,

- **constraints** (list,optional) constraints on fun,

- **bounds** (sequence,optional) bounds on x,

- **tol** (float,optional) tolerance for termination,

- **options** (dic,optional) method options,

- **callback** (callable, optional) function called after each iteration,

Some of these parameters are optional and may be omitted. This section provides examples of how some of the optimization functions described in the previous section can be used. The first optimization example uses the **minimize** function to optimize the volume of the box from Example 9.1. There is only a single variable and, therefore, it is a univariate function.

Example 9.2 Optimizing the volume of a box

As described in Example 9.1, the box example consists of maximizing the volume without having any constraints. As the first method to be used in the using the Nelder-Mead simplex algorithm indicating a method that minimizes a function without any constraints. The format to minimize the objective function **f** is:

```
res = minimize(f, x0, method = 'nelder-mead', options)
```

where **x0** is the initial condition and **options** is the set of options for the minimization process. For this example, the objective function is

$$Volume = -[4x^3 - 2(a + b)x^2 + abx]$$

which is defined with a $= 24$ and b $= 9$ in the following script as:

```
ab = [24, 9]
def volume(x, ab):
      return -(4*x**3-2*(ab[0] + ab[1])*x**2 + ab[0]*ab[1]*x)
# The sign of the function is changed to
# a minus sign to minimize.
```

The minimization instruction is:

```
res = minimize(volume, x0, args = ab, method = 'nelder-mead',
   options = {'xatol':1e-8,'disp':True})
```

The complete script is:

```
'''Script Example_9_2a.py
Obtaining the maximum volume for a box that is built
using a rectangular cardboard.'''

# Import the function minimize:
from scipy.optimize import minimize
x0 = 1 # initial value.
# Values for a and b
ab = [24, 9]

# Objective function:
def volume(x, ab):
      return -(4*x**3-2*(ab[0] + ab[1])*x**2 + ab[0]*ab[1]*x)
# The sign of the function is changed
# to a minus sign to minimize.

# Optimization begins:
res = minimize(volume, x0, args = ab, method = 'nelder-mead', \
            options = {'xatol':1e-8,'disp': True})

# Display the values of x and volume.
print('Optimum value for x: ', res.x[0])
print('Optimum volume: ', -res.fun)
```

The result is:

```
Optimization terminated successfully.
        Current function value: -200.000000
```

```
        Iterations: 31
        Function evaluations: 62
    Optimum value for x: 2.0000
    Optimum volume: 200.0
```

The first four lines are produced by the `disp = True` from `options`. The last two lines are due to the `print` instructions. The optimization process performed 31 iterations and 62 function evaluations.

In the minimization there were no derivatives computed. An improvement in the minimization, process can be achieved if a method that uses the Jacobian is used. In the case of an univariate function, the Jacobian is the first derivative. The method is the Broyden-Fletcher-Goldfarb-Shanno or simply BFGS. The derivative of the objective function is:

$$\frac{dVolume}{dx} = -(12x^2 - 4(a+b)x + ab)$$

The Python function that implements the Jacobian is:

```python
def jac0(x, ab):
    a, b = 24, 9
    return -(12*x**2 - 4*(a + b)*x + a*b)
```

The `minimize` instruction is:

```python
res = minimize(volume, x0, args = ab, method = 'BFGS', \
    jac = jac0, options = {'disp': True})
```

The complete script is:

```python
"'Script Example_9_2b.py
Obtaining the maximum volume for a box that is built
using a rectangular cardboard.
The method BFGS is used.'"

# Import the function minimize:
from scipy.optimize import minimize

x0 = 1 # initial value.
# Values for a and b
ab = [24, 9]
```

```
# Script continued from previous page.

# Objective function:
def volume(x, ab):
     return -(4*x**3-2*(ab[0] + ab[1])*x**2 + ab[0]*ab[1]*x)
# The sign of the function is changed to a
# minus sign to minimize.

# Jacobian.
def jac0(x, ab):
     return -(12*x**2-4*(ab[0] + ab[1])*x + ab[0]*ab[1])
# Optimization begins:
res = minimize(volume, x0, args = ab, method = 'BFGS', \
     jac = jac0, options = { 'disp': True})

# Display the values of x and volume.
print('Optimum value for x: ', res.x[0])
print('Optimum volume: ', -res.fun)
```

The result is:

```
Optimization terminated successfully.
Current function value: -200.000000
     Iterations: 4
     Function evaluations: 5
     Gradient evaluations: 5
Optimum value for x: 2.0000000003202487
Optimum volume: 200.0
```

The result is the same but the BFGS algorithm is faster by a factor of almost 8. The graph of volume versus x in Figure 9.2 shows that the volume is maximum when x = 2, as shown by the optimization process.

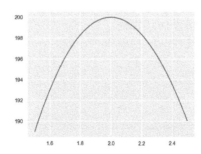

FIGURE 9.2: Graph of volume versus x.

Another improvement can be achieved in some cases by additionally using the Hessian. The Hessian of an univariate function f is the second derivative. The method is a Newton conjugate gradient abbreviated Newton-CG. The function of the Hessian in this example is:

$$\frac{dVolume}{dx} = -24x - 4(a + b)$$

that can be written as a function as:

```
def hes(x, ab):
    return -(24*x - 4*(ab[0] + ab[1]))
```

The complete script is

```
'''Script Example_9_2c.py
Obtaining the maximum volume for a box that is built
using a rectangular cardboard.
The method BFGS is used.'''

from scipy.optimize import minimize

# initial value.
x0 = 1

# Values for a and b
ab = [24, 9]

# Objective function. The sign of the function
# is changed to -v to minimize.
def volume(x, ab):
    return -(4*x**3-2*(ab[0] + ab[1])*x**2 + ab[0]*ab[1]*x)

# Jacobian.
def jac0(x, ab):
    return -(12*x**2-4*(ab[0] + ab[1])*x + ab[0]*ab[1])

# Hessian.
def hes(x, ab):
    return -(24*x - 4*(ab[0] + ab[1]))
```

```
# Script continued from previous page.

# Optimization begins. v is the volume.
res = minimize(volume, x0, args = ab, method = 'Newton-CG', \
     jac = jac0, options = {'xtol': 1e-8, 'disp': True})

# Display the values of x and volume.
print(res.x)
print(res.fun)
```

The result is:

```
Optimization terminated successfully.
Current function value: -200.000000
        Iterations: 5
        Function evaluations: 6
        Gradient evaluations: 11
        Hessian evaluations: 0
[2.00]
-200.0
```

The results are similar to the previous script. But in some other functions, especially in multivariate functions, the results may be much better.

Example 9.3 Optimization of a multivariate function
Given the multivariate objective function

$$f(x) = -(2xy + 2x - x^2 - 2y^2)$$

subject to the constraints:

$$x^3 - y = 0$$
$$y - (x - 1)^4 - 2 \leq 0$$

and bounds

$$0.5 \leq x \leq 1.5$$
$$1.5 \leq y \leq 2.5$$

The objective function, with $x = $ x[0] and $y = $ x[1] is

```
def f(x):
        return -(2*x[0]*x[1] + 2*x[0] - x[0]**2 - 2*x[1]**2)
```

The Jacobian is a list given by

$$J(x) = [-(2y + 2 - 2x), -(2x - 4y)]$$

In Python this is implemented by

```
def jc(x):
    return [-(2*x[1] + 2 - 2*x[0]), - (2*x[0] - 4*x[1])]
```

With respect to the constraints, there is an equality constraint

$$x^3 - y = 0$$

which can be implemented by

```
def consf(x):
    return np.array([x[0]**3 - x[1]])
```

with Jacobian

$$[3x^2, -1]$$

which is implemented by

```
def jac(x):
    return np.array([3.0*(x[0]**2.0), -1.0])
```

In addition, there is an inequality constraint

$$y - (x - 1)^4 - 2 \leq 0$$

which is implemented by

```
def funIn(x):
    return np.array([x[1] - (x[0]-1)**4 - 2])
```

These constraints are

```
cons = ({'type': 'eq', 'fun' : consf, 'jac' : jcc},
{'type': 'ineq', 'fun' : funIn})
```

The functions defined above only have an instruction for the definition of them; therefore, a lambda function may be used, as:

```
cons = ({'type': 'eq',
        'fun' : lambda x: np.array([x[0]**3 - x[1]])},
```

```
{'jac' : lambda x: np.array([3.0*(x[0]**2.0), -1.0])},
{'type': 'ineq',
 'fun' : lambda x: np.array([x[1] - (x[0]-1)**4 - 2])})
```

The bounds and the initial values x0 are:

```
bnds = ((0.5, 1.5), (1.5, 2.5))
x0 = [0, 2.5]
```

The function `minimize` is used with the bounds and initial value. First, the constraints are not used, and then they are included:

```
result = opt.minimize(f, x0, constraints = None)
```

The complete script is:

```
# Script Example_9_3.py
import numpy as np
from scipy.optimize import minimize
import scipy.optimize as opt

# Bounds and initial values
bnds = ((0.5, 1.5), (1.5, 2.5))
x0 = [0, 2.5]

# Objective function
def f(x):
        return -(2*x[0]*x[1] + 2*x[0] - x[0]**2 - 2*x[1]**2)

# Constraints
cons = ({'type': 'eq',
        'fun' : lambda x: np.array([x[0]**3 - x[1]]),
        'jac' : lambda x: np.array([3.0*(x[0]**2.0), -1.0]),
        {'type': 'ineq',
        'fun' : lambda x: np.array([x[1] - (x[0]-1)**4 - 2])})

# Optimization function minimize
result = opt.minimize(f, x0, constraints = None)
print(result)
```

to obtain:

```
message: 'Optimization terminated successfully.'
success: True
 status: 0
```

```
       fun: -1.9999999999996365
         x: [2.000e+00    1.000e+00]
       nit: 5
       jac: [ 1.252e-06     -1.416e-06]
 hess_inv: [[9.983e-01     5.011e-01]
           [5.011e-01     4.994e-01 ]]
      nfev: 18
      njev: 6
```

Now the `minimize` instruction row is replaced by

```
result = opt.minimize(f, x0, bounds = bnds, constraints = cons)
```

and the result is:

```
 message: 'Optimization terminated successfully.'
 success: True
  status: 0
     fun: 2.049915472024102
       x: [1.26089314, 2.00463288]
     nit: 6
     jac: [-3.487e+00     5.497e+00]
    nfev: 19
    njev: 6
```

9.6   Linear Programming

Linear programming is an optimization technique used when the objective function only has linear terms. This technique was developed during the World War II. The objective function is of the form:

$$c_1 x_1 + c_2 x_2 + \cdots + c_n x_n = z \tag{9.1}$$

Subject to the constraints:

$$\begin{cases} a_{11}x_1 + a_{12}x_2 + \cdots + a_{1n}x_n & \leq b_1 \\ a_{21}x_1 + a_{22}x_2 + \cdots + a_{2n}x_n & \leq b_2 \\ \vdots \\ a_{m1}x_1 + a_{m2}x_2 + \cdots + a_{mn}x_n & \leq b_m \\ x_1, x_2, \cdots, x_n & \geq 0 \end{cases} \tag{9.2}$$

The variables

$$x_1, x_2, \cdots, x_n$$

are called decision variables. The set of decision variables that satisfy the constraints is called a feasible point. The set of every possible feasible point is called the feasible region. In the constraint equations, some of the inequalities can be equalities; thus, there is a set of inequality constraints and a set of equality constraints.

SciPy includes a function to perform linear programming. The function is linprog with the following format:

```
from scipy.optimize import linprog
linprog(f, A_ub, b_ub, A_eq, b_eq, bounds, method, options, x0)
```

where the parameters are

- f is an array with the coefficients of the objective function.

- A_ub is the array with the coefficients of the inequality constraints.

- b_ub is the array with the inequality constraints.

- A_eq is the array with the coefficients of the equality constraints.

- b_eq is the array with the equality constraints.

- method The methods supported are 'highs-ds', 'highs-ipm', 'highs', 'interior-point' (default), 'revised simplex', and 'simplex'.

- bounds are the lower and upper bounds for each variable x.

- options is a dictionary with: the maximum number of iterations maxiter, the display value disp set to True to display the results.

- tol the tolerance to determine if the solution is within the specified range.

- x0 the initial values for x.

The process for linear programming optimization is shown with an example.

Example 9.4 Linear programming
It is desired to minimize the function

$$f = 40x + 60y$$

subject to constraints

$$\begin{bmatrix} 2 & 1 \\ 1 & 1 \\ 1 & 3 \end{bmatrix} \begin{bmatrix} x \\ y \end{bmatrix} \leq \begin{bmatrix} 70 \\ 40 \\ 90 \end{bmatrix}$$

$$x \geq 0, y \geq 0$$

The parameters are

```
f = [-40, -60]
A_ub = [[2, 1, [1, 1], [1, 3]]
b_ub = [70, 40, 90]
bnds = [(0, None), (0, None)]
```

The sign of f is changed to minimize. The `linprog` method is

```
res = linprog(f, A_ub, b_ub, bounds = bnds)
```

Finally, the results are printed:

```
print(res)
```

The complete script is:

```
"' This is script Example_9_4a.py
This program implements linear programming.'"

# import the function linprog:
from scipy.optimize import linprog

# Coefficients of the objective function:
f = [-40, -60]

# Arrays of the inequality constraints:
A_ub = [[2, 1], [1, 1], [1, 3]]
b_ub = [70, 40, 90]

# bounds:
bnds = [(0, 100), (0, 100)]

# linprog method
res = linprog(f, A_ub, b_ub, bounds = bnds)

# Results are printed:

print(res)
```

The results are:

```
     con: array([ ], dtype = float64)
     fun: -2099.9999997905556
 message: 'Optimization terminated successfully.'
     nit: 3
   slack: array([15., 0., 0.])
  status: 0
 success: True
       x: array([15., 25.])
```

The optimal values for x are x0 = 15, x1 = 25. The optimal value for the objective function is 2100. The optimization process made 3 iterations given by nit.

This same problem can be solved with CVXOPT using the solvers.lp method. The only requirement is to convert the objective function, the matrix A_ub and the vector b_ub to numpy arrays and cvxopt matrices with floating-point elements. They can be converted to floating-point with the option, tc = 'd'. The library cvxopt and its methods solvers and matrix have to be imported. Then, the final script using the library CVXOPT is:

```
"' This is script Example_9_4b.py
This program realizes linear programming.
It uses the library CVXOPT.'"

# Import the necessary libraries and methods:
from cvxopt import solvers, matrix
import numpy as np

# Coefficients of the objective function:
f = matrix(np.array([-40.0, -60.0]), tc = 'd')

# Arrays of the inequality constraints:
# The upper and lower bounds are incorporated in A_ub.
A_ub=matrix(np.array([[2,1],[1,1],[1,3],[1,0],[0,1]]),tc='d')
b_ub = matrix(np.array([70, 40, 90, 100, 100]), tc = 'd')

# solvers.qp method
res = solvers.lp(f, A_ub, b_ub)
```

```
# Script continued from previous page.

# Results are printed:
x = res['x']
ObjFun = res['primal objective']
print('The values of x and y are: ')
print('The value of x is: ', x[0])
print('The value of y is: ', x[1])
print('The value of the objective function is: ', ObjFun)
```

If this program is executed the results are:

```
Optimal solution found.
The values of x and y are:
The value of x is: 15.000000511662899
The value of y is: 25.000000067714527
The value of the objective function is: -2100.0000245293877
```

which are the same as obtained with `linprog`.

There is another package for linear programming with the name `pulp`. There are some instructions that allows users to create new variables, and new problems, combine variables, use a solver, display the status of the solution, to get the values of the variables. Consider the linear programming example from above. To solve that problem with `pulp` first import the `pulp` library

```
import pulp
```

Then define the linear programming problem

```
linearProblem = pulp.LpProblem('Maximizing objective function',
pulp.LpMaximize)
```

Now the variables are defined as

```
x1 = pulp.LpVariable('x1', lowBound = 0)
x2 = pulp.LpVariable('x2', lowBound = 0)
```

The objective function is defined as

```
linearProblem += 40*x1 + 60*x2
```

The linear constraints are given as

```
linearProblem += 2*x1 + x2 <= 70
```

```
linearProblem += x1 + x2 <= 40
linearProblem += x1 + 3*x2 <= 90
```

The linear programing problem is solved with

```
solution = linearProblem.solve()
```

Finally, the results of the optimization problem are displayed as

```
print(str(pulp.LpStatus[solution]) + \
" ; max value = " + str(pulp.value(linearProblem.objective)))
print("x1_opt =" + str(pulp.value(x1)))
print("x2_opt =" + str(pulp.value(x2)))
```

The results are displayed as:

```
Optimal max value = 2100.0
x1_opt = 15.0
x2_opt = 25.0
```

which are the same values obtained with the other two packages. The complete script is

```
# File Example_9_4c.py
# Solves the linear programming with pulp.

# Import pulp
import pulp

# The linear programming problem:
linearProblem = pulp.LpProblem('Maximizing objective function' \
    , pulp.LpMaximize)

# The variables are defined as:
x1 = pulp.LpVariable('x1', lowBound = 0)
x2 = pulp.LpVariable('x2', lowBound = 0)

# The objective function is:
linearProblem += 40*x1 + 60*x2

# The linear constraints are:
linearProblem += 2*x1 + x2 <= 70
linearProblem += x1 + x2 <= 40
linearProblem += x1 + 3*x2 <= 90
```

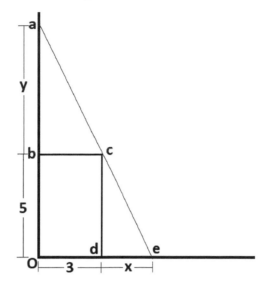

FIGURE 9.3: Problem about a ladder.

```
# Script continued from previous page.

# The linear programing problem is solved with:
solution = linearProblem.solve()

# The results are:
print(str(pulp.LpStatus[solution]) + \
" ; max value = " + str(pulp.value(linearProblem.objective))
print("x1_opt =" + str(pulp.value(x1)))
print("x2_opt =" + str(pulp.value(x2)))
```

Example 9.5 Problem about a ladder

The problem with a ladder involves placing a ladder leaning against a wall of height 5 m so that it touches the wall and its length L is at a minimum as shown in Figure 9.3. To obtain the equations to be solved it can be seen that the largest triangle must satisfy the Pythagorean theorem. Therefore,

$$(x + 3)^2 + (y + 5)^2 = L^2$$

This equation corresponds to a circle centered at (-3, -5) and radius L. This would be the target function and, therefore, this function should be minimized.

For the constraints, the small triangles abc and cde are considered. These triangles are similar and they are related by:

$$\frac{x}{5} = \frac{3}{y}$$

which it is equivalent to:

$$xy = 15$$

From this relationship, it is clearly seen that y is dependent upon x. Thus, there is only an independent variable. Substituting this relationship in the equation for the ladder, this equation becomes:

$$(x + 3)^2 + (15/x + 5)^2 = L^2$$

and the length of the ladder L has to be minimized.
    Then, the file for the objective function is:

```
# Objective function:
def ladder(x):
    return (x + 3)**2 + (5 + 15/x)**2
```

The complete script to perform optimization is:

```
# File Example_9_5.py
# Obtains the values of x, and the problem about the ladder.

# Import the required libraries and functions:
from scipy.optimize import minimize_scalar
import numpy as np
import matplotlib.pyplot as plt

# Objective function:
def ladder(x):
    return (x + 3)**2 + (5 + 15/x)**2

# Initial condition and bounds:
x0 = 1
bounds = (0, 100)

# Implement the optimization:
res = minimize_scalar(ladder, x0, bounds,\
        options = {'disp':True}, method = 'bounded')
```

```
# Script continued from previous page.

# Print out the results:
x = res.x
L = np.sqrt((x + 3)**2 + (15/x + 5)**2)

print("The optimum value for is ", x)
print("The optimum value for L is ", L)

# Plotting the ladder equation and the optimum value:
xp = np.linspace(4, 4.43, 100) # Range for the plot.
Lp = np.sqrt((xp + 3)**2 + (5 + 15/xp)**2) # Ladder equation.
Lmin = np.sqrt((x + 3)**2 + (5 + 15/x)**2) # Optimum value.
# Plot minimum:
plt.text(x, Lmin, 'o' , size = 15, va = 'center', ha = 'center')
plt.plot(xp, Lp) # Plot the ladder equation.
plt.grid(True, linestyle = '-.')
plt.show()
```

The results are:

```
Optimization terminated successfully.
The returned value satisfies the termination criteria
(using xtol = 1e-05 )
The optimum value for x is 4.217161887614067
The optimum value for L is 11.194099763086815
The optimum value for the function is 125.30786950594029
```

This file provides the values of x to optimize the problem and additionally calculates the length of the ladder. At the end, the curve of the objective function is plotted. The result of the minimization gives:

```
L = 11.1941
```

The graph for the function **ladder** is shown in Figure 9.4. Here, it can clearly be seen that the value of x minimizes the objective function and is consistent with the value resulting from the optimization.

There are many ways to solve this example. This is only a solution but the reader is invited to try an other optimization algorithm.

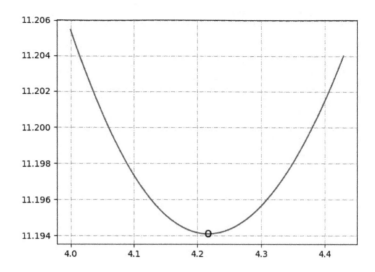

FIGURE 9.4: Plot of the objective function with the minimum value displayed.

9.7   Quadratic Programming

There are several non-linear problems that can be solved with optimization techniques. These non-linear problems are usually multi-variate. An important case is when the objective function to minimize is of the form:

$$\frac{1}{2}x^T A x + p^T x$$

subject to $Bx \leq b$

$$Cx = c$$

where x is an n-dimensional variable vector, x^T is the transpose of x, A and B are $n \times n$-dimensional symmetric matrices, p is a vector of coefficients and p^T is its transpose, and b and c are n-dimensional real vectors. If P is positive-semidefinite, then the objective function is convex and the minimum obtained is a global minimum.

For optimization, `Scipy` does not have a function that minimizes quadratic programming objective functions. However, there exist two other libraries that have specific functions for quadratic programming optimization. They are the libraries `CVXOPT` and `solvers.qp`. The next example is solved with `CVXOPT`.

Example 9.6 Quadratic programming
To minimize the quadratic form

$$\frac{1}{2}x^2 + 3x + 4y$$

with constraints

$$x \geq 0$$
$$y \geq 0$$
$$x, y \geq 0$$
$$x + 3y \geq 15$$
$$2x + 5y \leq 100$$
$$3x + 4y \leq 80$$

The quadratic form can be written in the standard form shown above as:

$$\frac{1}{2} \begin{bmatrix} x & y \end{bmatrix} \begin{bmatrix} 1 & 0 \\ 0 & 0 \end{bmatrix} \begin{bmatrix} x \\ y \end{bmatrix} + \begin{bmatrix} 3 & 4 \end{bmatrix} \begin{bmatrix} x \\ y \end{bmatrix}$$

Then, the vectors and matrices are given by:

$$x = \begin{bmatrix} x \\ y \end{bmatrix}, A = \begin{bmatrix} 1 & 0 \\ 0 & 0 \end{bmatrix}, p = \begin{bmatrix} 3 \\ 4 \end{bmatrix}, B = \begin{bmatrix} -1 & 0 \\ 0 & -1 \\ -1 & -3 \\ 2 & 5 \\ 3 & 4 \end{bmatrix}, b = \begin{bmatrix} 0 \\ 0 \\ -15 \\ 100 \\ 80 \end{bmatrix}$$

These matrices and vectors in Python, converted to numpy arrays and `CVXOPT` matrices are (`numpy` and `cvxopt` have to be imported previously):

```
A = matrix(np.array([[1.0, 0.0], [0.0, 0.0]]))
p = matrix(np.array([3.0, 4.0]))
B = matrix(np.array([[-1.0,0],[0,-1.0],[-1.0,-3.0], \
                 [2.0,5.0],[3.0,4.0]]))
b = matrix(np.array([0.0, 0.0, -15.0, 100.0, 80.0]))
```

The matrices and vectors are in floating-point format. This is because `CVXOPT` only works with floating-point data.

To perform the minimization, the required functions have to be imported. The function from CVXOPT for quadratic programming optimization is solvers.qp. This function is imported with:

```
from cvxopt import solvers
```

Then the minimization is done with:

```
solution = solvers.qp(A, p, B, b)
```

The solver displays information about the process but the important information is the value of the objective function and the parameters x and y. These are displayed with:

```
x = solution['x']
ObjFun = solution['primal objective']
print('The values of x and y are: ')
print('The value of x is: ', x[0])
print('The value of y is: ', x[1])
print('The value of the objective function is: ', ObjFun)
```

The complete script is:

```
"' This is script Example_9_6.py
This program minimizes the function 0.5x**2 + 3x + 4y.
It uses the quadratic programming solver from CVXOPT.'"

# Import the solvers from library cvxopt:
from cvxopt import solvers, matrix
import numpy as np

# Define the vectors and matrices:
A = matrix(np.array([[1.0, 0.0], [0.0, 0.0]]))
p = matrix(np.array([3.0, 4.0]))
B = matrix(np.array([[-1.0,0],[0,-1.0],[-1.0,-3.0],
                     [2.0,5.0],[3.0,4.0]]))
b = matrix(np.array([0.0, 0.0, -15.0, 100.0, 80.0]))

# The minimization is done with:
solution = solvers.qp(A, p, B, b)

# The solution and the objective function are displayed:
x = solution['x']
ObjFun = solution['primal objective']
```

```
# Script continued from previous page.

# The solution and the objective function are displayed:
print('The values of x and y are: ')
print('The value of x is: ', x[0])
print('The value of y is: ', x[1])
print('The value of the objective function is: ', ObjFun)
```

After the program is run the solution obtained is:

```
Optimal solution found.
The values of x and y are:
The value of x is: 7.131816408857141e-07
The value of y is: 5.000001008391809
The value of the objective function is: 20.00000617311241
```

Thus, it can be seen that the values for x and y are 0 and 5, respectively. The value of the objective function is the value of the quadratic form which in this case is 20.0.

9.8   Python Instructions for Chapter 9

Table 9.1 lists the instructions in Chapter 9

TABLE 9.1:   Python instructions for Chapter 9

Instruction	Description
scipy	Library of mathematical functions.
minimize	Method to minimize a function.
nelder-mead	Method for optimization.
optimize	Library of optimization methods.
Newton-CG	Method for optimization.
Jacobian	Jacobian array.
hess_inv	Hessian matrix inverter.
jac	Obtains the Jacobian array.
linprog	Used for linear programming optimization.
CVXOPT	Library of optimization functions.
solvers	Library of optimization functions. Part of cvxopt.
minimize_scalar	Method to minimize scalar functions.
qp	Method to optimize quadratic forms.
BFGS	Method for minimization.

9.9 Conclusions

A brief introduction to the topic of optimization using Python and two of the optimization packages available, `SciPy` and `CVXOPT`, have been presented. Optimization using linear and quadratic programming as well as minimization with and without constraints has been implemented. Several examples have shown how the optimization process is done.

9.10 Selected Bibliography

[1] Optimization Toolbox User's Guide, The Mathworks, Inc., Natick, MA, 2018.

[2] P. Venkataraman, Applied Optimization with MATLAB Programming, J. Wiley and Sons, New York, 2002.

Chapter 10

Image Processing with OpenCV

10.1 Introduction

A black and white image is a two-dimensional representation of three-dimensional information in color. For processing, an image is partitioned into a number of elements called pixels. This word comes from the combination of the English words PICture ELement. The standard size of images in modern devices is in the range of megapixels. In Figure 10.1, it is shown how an image is represented. From this figure it can be seen that there are $8 \times 8 = 64$ pixels. A natural way of representing an image is then by a matrix or array where each pixel position is associated with the values n and m of the matrix elements. The value of the element m, n of the matrix is then the value of the gray level of the corresponding pixel.

The way to number the pixels is exactly equal to the representation of matrices as shown in Figure 10.2. Thus, the 0,0 pixel is a white pixel and pixel 3,4 is a black pixel. Each pixel has an intensity level. For monochrome images, the levels of intensity are gray levels. Depending on how many bits are handled for this representation, it will be the number of gray levels that can be represented. The image of Figure 10.2 is a chessboard that only has two levels and therefore can be represented by a single bit. For example, if

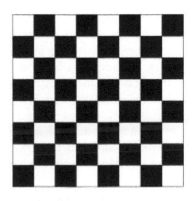

FIGURE 10.1: An image.

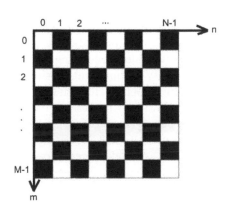

FIGURE 10.2: Axes for images.

DOI: 10.1201/9781003222118-10

FIGURE 10.3: Lena image.

that bit is 0 then it is black and if the pixel is 1 it is white, and vice versa. For gray level images to be represented adequately, at least 256 gray levels are needed which requires that each pixel is represented with 8-bit words. An 8-bit word is called a byte. In the case of color images, there are three color components; then, 24 bits, or three bytes, are required for each color pixel.[1]

OpenCV is a software package especially designed for image and video processing. It was initially developed by Intel and since August 2012 support for OpenCV was taken by the non-profit foundation OpenCV.org which provides a developer and user site. It is written in C++ and has an interface for Python. This chapter provides an introduction to image and video processing using OpenCV and presents a set of instructions that allow users to begin to do image and video processing. There are many more instructions available in the OpenCV documentation site `https://docs.opencv.org`. Instructions for installing OpenCV in different platforms are available in the Appendix. OpenCV allows the processing of images in different formats, such as `bmp`, `eps`, `tif`, `pcx`, `gif`, `png`, etc. and even in a free format. To test our scripts in Python using OpenCV, an image known as Lena is used. This image is usually used as a reference by researchers. This image of Lena is shown in Figure 10.3. The original image is chromatic, in a `bmp` format and has a size of 512×512×3.

---

[1]Image files that are discussed in this chapter can be found in the web page of the book.

TABLE 10.1: Flags for `imread`.

Flag	Action	Value
Nothing	It reads the unchanged image.	-1
IMREAD_UNCHANGED	It reads the unchanged image.	-1
cv.IMREAD_GRAYSCALE	It converts the image to a single channel grayscale image.	0
cv.IMREAD_COLOR	It converts the image to a 3 channel BGR color image.	1

10.2 Reading and Writing Images and Videos

The first step for using OpenCV is to import the libraries with

```
import cv2
```

This step is needed to have access to all the functions available for image and video processing. The name `cv2` is the alias for OpenCV. The instruction to read an image is

```
file_name = cv2.imread("image", flags)
```

where `image` specifies the path and the name for the image file. The flags are shown in Table 10.1. The image is read into an image file with the name `file_name`. To display this image the instruction `imshow` is used. The format is

```
cv2.imshow("window_name", file_name)
```

where `window_name` is the name in the window which displays the image. An instruction `cv2.waitKey(0)` can be added to have the image windows open until the key `Enter` is pressed and the pointer is on the image. Then, the instruction `cv2.destroyAllWindows()` closes the windows. It is possible to write a number inside the parenthesis, for example, `cv2.waitKey(5000)` where the number is the time in milliseconds that the image window remains open.

Example 10.1 Reading and displaying an image
As an example, the image `lena.jpg` stored in the folder `images` can be read with:

```
Lena = cv2.imread("images/lena.jpg", cv.IMREAD_COLOR)
```

Now the image `lena.jpg` is in the OpenCV image file **Lena** as a color image. To display the image, the instructions are:

```
imshow("window name", Lena)
waitKey(0)
destroyAllWindows()
```

where `window name` is the name of the window displayed and `image` is the image name. The instruction `waitKey(0)` mantains the window open until any key is pressed. Then, all image windows are closed by the instruction `destroyAllWindows()`. The following script is used to read and display the image **Lena**:

```
"' This is script Example 10_1.py.
This program reads an image from a file.'"

# The necessary library is imported:
import cv2

# The image Lena is read:
Lena = cv2.imread("images/lena.jpg")

# The image is displayed:
cv2.imshow("Lena", Lena)
cv2.waitKey(0)
cv2.destroyAllWindows()
```

The method `waitKey(0)` is waiting for an input indefinitely. The image displayed closes when the pointer is placed upon the image and any key is pressed. If another number different from 0 is entered, it indicates the number of milliseconds that the image is displayed. The method `destroyAllWindows()` then closes every open window.

    The image **Lena** can be saved to a different format such as **jpg** or **png**, for example. The instruction is `imwrite(fileName, image)`. For this example, to save the image **Lena** in format **png**, `imwrite` is used as

```
cv2.imwrite('images/lena.png', Lena)
```

Now, after closing the image window, the folder **images** contains the images `lena.jpg` and `lena.png`. In the case of **jpg** and **png** images, the instruction `imwrite` can use additional parameters like the quality of the image saved.

FIGURE 10.4: Lena image.   FIGURE 10.5: Low quality Lena.

A list of the parameters can be found at the opencv.org site[2]. For example, for the lena.jpg image in Figure 10.4 the parameter cv2.IMWRITE_JPEG_QUALITY can be changed from 0 to 100 (100 is the better quality) and this indicates the level of compression in the image. In order to have a larger compression, the instruction used is

```
cv2.imwrite('images/lenaLow.jpg, ' \
            Lena, [(cv2.IMWRITE_JPEG_QUALITY), 5])
```

The resulting image, shown in Figure 10.5 is compared to the original image and it can clearly be seen that the image with a large compression has lost a great deal of details. The complete script to read the lena.jpg image and write it in a jpg format with 5% quality is:

```
"' This script Example 10_1b.py.
This program reads an image and saves it to a file in format
jpg with 5% quality"

# The necessary library is imported:
import cv2

# The image Lena is read:
Lena = cv2.imread("images/lena.jpg")

# Script continues at next page.
```

[2]https://docs.opencv.org/master/d8/d6a/group__imgcodecs__flags.html#ga292d 81be8d76901bff7988d18d2b42ac

```
# Script continued from previous page.

# The image is saved with a high compression of 5%:
cv2.imwrite('images/lenaLow.jpg, ' \
            Lena, [(cv2.IMWRITE_JPEG_QUALITY), 5])

# LenaLow is read as an OpenCV image:
LenaLow = cv2.imread("images/lenaLow.jpg")

# The images are displayed:
cv2.imshow("Lena high quality", Lena)
cv2.imshow("Lena low quality", LenaLow)
cv2.waitKey(0)
cv2.destroyAllWindows()
```

10.2.1   Image Representation in Python

An image is represented in Python as a matrix. In a (8-bit) greyscale image each picture element has an assigned intensity that ranges from 0 to 255. A gray-scale image is a black-and-white image and includes many shades of gray. Each pixel is represented by a number which for an 8-bit image representation has 256 values going from 0 for a black pixel to 255 for a white one. The pixels are arranged in a matrix structure that has rows and columns. The number of rows and columns determines the size of the image.

The image size can be determined with **shape** which is a property of the image. For example, for the image Lena this can be done as:

```
a = lena.shape
print(a)
```

which in this image is

```
(512, 512, 3)
```

These numbers mean that the image is represented by an array that has 512 rows, 512 columns, and has three layers corresponding to the three colors BGR. A portion of an image can be cropped using the fact that images are represented as matrices and taking into account that the first argument refers to the height as:

```
image[y1:y1 + h, x1:x1 + k]
```

where the region of interest (ROI) is the square with vertices at

```
(x1, y1), (x1 + k-1, y1), (x1, y1 + h-1), (x1 + k-1, y1 + h-1)
```

FIGURE 10.6: Image cropped.

Consider the Lena image. A script to crop it is:

```
"' This is script Crop_image.py.
This program crops an image"'

# The necessary libraries are imported:
import cv2

# The image Lena is read:
Lena = cv2.imread('images\lena.jpg')
print('Lena', Lena.shape)

# The image is cropped:
LenaCrop = Lena[0:300, 100:450]
print('Lena Cropped', LenaCrop.shape)

# The images are displayed:
cv2.imshow('Lena', Lena)
cv2.imshow('Lena cropped', LenaCrop)
cv2.waitKey(0)
cv2.destroyAllWindows()
```

The cropped image is shown in Figure 10.6.

The most used format in color images is the RGB format. The initials RGB correspond to the colors red, green, and blue. For each color, there is a matrix or layer, which means that the first layer is the red layer, the second one is the green layer, and the last one is the blue layer. However, in OpenCV the layers are arranged as BGR. That is, the color layers are inverted. When an image is opened with OpenCV as in the examples above, images are displayed in the correct form. However, when an image is displayed with `matplotlib`, then the colors are changed because the color formats are different.

The conversion from RGB image format to another one can be done when the image is read using `imread`. For example, to read the image as a gray format image use:

```
cv2.imread("image", cv2.RGB2GRAY)
```

If the image has already been read, then the instruction is `cvtColor` as:

```
Output_Image = cv2.cvtColor(input_Image, cv2.COLOR_BGR2GRAY)
```

Another color conversion is changing from RGB to HSV format. HSV stands for hue, saturation, and chroma. This color space is used in a number of applications. The instruction is:

```
Output_Image = cv2.cvtColor(input_Image, cv2.COLOR_BGR2HSV)
```

```
''' This is script Format_conversion.py.
This program changes the format of an image'''

# The necessary libraries are imported:
import cv2

# The image Lena is read:
Lena = cv2.imread('images\lena.jpg')

# The image format is changed to HSV:
LenaHSV = cv2.cvtColor(images\lena.jpg, cv2.COLOR_BGR2HSV)

# The layers of the HSV image are obtained:
h, s, v = cv2.split(LenaHSV)

cv2.imshow('Lena', Lena)
cv2.imshow('Lena H', h)
cv2.imshow('Lena s', s)
cv2.imshow('Lena V', v)
cv2.waitKey(0)
cv2.destroyAllWindows()
```

10.3 Video Capture and Display

The process to read and capture videos is similar to that of images. For videos what is captured are video frames. The first step is to indicate what is the source of the video. In the first example, the computer webcam is used to capture the video frames and this is the default video source. The instruction `capture = cv2.VideoCapture(0)` indicates that the computer web cam, which is indicated by the 0 inside the parentheses, is being used to capture frames. Other cams available are designated with a different number. The actual reading of the frames is done with the instruction `capture.read()` with the following format:

```
correct, frame = capture.read()
```

The variable `correct` is a boolean variable which with a `True` value indicates that the frames are being captured correctly. The variable `frame` contains the frames and to read them a `while` loop is used as follows:

```
while True:
    correct, frame = capture.read()
    cv2.imshow('Frame', frame)
```

The `imshow` instruction is used to display the frames as they are captured. To exit the `while` loop the instruction `waitKey` is used together with any key pressed, for example the letter q to quit the capture. A condition `if` is used as:

```
if cv2.waitKey(1) & 0xFF == ord('q'):
    break
```

Finally, the instructions needed to release the web cam are:

```
capture.release()
cv2.destroyAllWindows()
```

The complete file to capture video from the computer web cam is

```
"' This is script Video_capture.py
This program captures a video stream from the computer web
cam.'"


# The necessary library is imported:
import cv2


# Establish the video source.
capture = cv2.VideoCapture(0)


# With a while loop capture the frames:
while True:
    correct, frame = capture.read()
    cv2.imshow('Frame', frame)


# Establish the condition to stop video capture.
# The key corresponding to the letter q is chosen.
    if cv2.waitKey(1) & 0xFF == ord('q'):
        break


# The instructions needed to release the web cam are:
capture.release()
cv2.destroyAllWindows()
```

A video can also be captured specifying the path and the video file name.

10.3.1 Saving a Video File

To save a sequence of frames to a file it is needed to specify the file name, the codec for the video capture from the FOURCC list, the number of frames per second, the frame size, and the color flag which is True for color frames. FOURCC is a 4-byte code. For Windows the XVID codec is usually used. For OSX the most used ones are mp4v for mp4 videos, DIVX for avi, and X264 for mkv. For this example run in Windows, the frames for the web cam are used and saved with:

```
fourcc = cv.VideoWriter_fourcc(*'XVID')
out = cv.VideoWriter('output.avi', fourcc, 20.0, (640, 480))
```

These instructions should be placed inside the `while` loop. The complete file is:

```
"' This is script Video_stream_capture.
This program captures a video stream from a video file.'"

import cv2 as cv
cap = cv.VideoCapture(0)

# Define the codec and create VideoWriter object
fourcc = cv.VideoWriter_fourcc(*'XVID')
out = cv.VideoWriter('output.mp4', fourcc, 20.0, (640, 480))
while True:
    ret, frame = cap.read()

    # write the frame.
    out.write(frame)
    cv.imshow('frame', frame)
    if cv.waitKey(1) == ord('q'):
        break
# Release everything.
cap.release()
out.release()
cv.destroyAllWindows()
```

The saved video in mp4 format is saved to file `output.mp4`.

10.4   Binary Images

An image can be binarized; that is, each one of the pixels can be made a 0 or a 1, which is equivalent to 0 (black) and 255 (white) in 8-bit images. The process to binarize an image is known as thresholding. It is made in grayscale images and if the pixel value is below a certain threshold that pixel is made 0. On the opposite side, if the pixel value is larger than this threshold, then the pixel value is set at 255. The instruction for thresholding in OpenCV is

```
cv2.threshold(Source, Value, max_Val, Technique)
```

where the parameters are:

- `Source`: Input Image array (must be in Grayscale).

- `Value`: Value of threshold where pixel values will change.

- `maxVal`: Maximum value that can be assigned to a pixel.

- `Technique`: Type of thresholding to be applied.

The available techniques are:

1. `cv2.THRESH_BINARY`: If pixel intensity is greater than the set threshold, value set to 255, else set to 0 (black).

2. `cv2.THRESH_BINARY_INV`: Opposite case of `cv2.THRESH_BINARY`.

3. `cv.THRESH_TRUNC`: If the pixel intensity value is greater than the threshold, it is truncated to the threshold. The pixel values are set to be the same as the threshold. All other values remain the same.

4. `cv2.THRESH_TOZERO`: Pixel intensity is set to 0, for all the pixels intensity, less than the threshold value.

5. `cv2.THRESH_TOZERO_INV`: Opposite case of `cv2.THRESH_TOZERO`.

Example 10.2 Binarization of an image

Consider the Lena image. If the binarization is made considering that if a pixel has a value of 128 then its value is set to 0, but if the value is above this threshold then its value is set to 255. The following script makes this thresholding:

```
"' This is script Example_10_2.py
This program binarizes an image."'

# Import OpenCV
import cv2

# Path to input image is specified and
# the image is loaded with imread command
Lena = cv2.imread('lena.jpg')

# cv2.cvtColor is applied over the
# image input with applied parameters
# to convert the image in grayscale
img = cv2.cvtColor(Lena, cv2.COLOR_BGR2GRAY)

# Applying two different thresholding
# techniques on the input image
# all pixels value above 128 will
# be set to 255 (white) in the first thresholding,
# and for the second thresholding technique
# all pixels above 60 will be set to 255 (white).

# Script continues at next page.
```

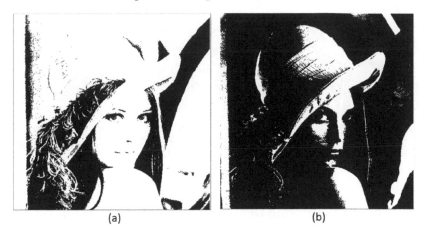

(a)   (b)

FIGURE 10.7: a) Binary threshold, b) Binary threshold inverted.

```
# Script continued from previous page.

ret, thresh1 = cv2.threshold(img, 60, 255,
cv2.THRESH_BINARY)
ret, thresh2 = cv2.threshold(img, 128, 255,
cv2.THRESH_BINARY)
# Windows showing output images
# with the thresholding technique
# applied to the input image.
cv2.imshow('Binary Threshold', thresh1)
cv2.imshow('Binary Threshold Inverted', thresh2)

# Delete images displayed.
cv2.waitKey(0)
cv2.destroyAllWindows()
```

The resulting images are shown in Figure 10.7.

Another more interesting binarization is the adaptive one. This type of binarization may be useful in cases where the image has different levels of illumination. There are two techniques available in OpenCV, in both cases the format is:

`adaptiveThreshold(in, out, max, Technique, thresType, block, C)`

where

- in is the original image.

- out is the resulting image.

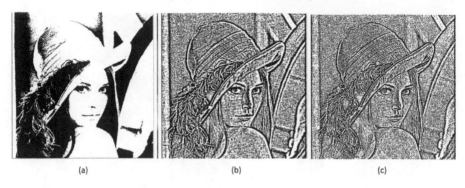

(a)   (b)   (c)

FIGURE 10.8: a) Binary threshold, b) Mean C threshold, and c) Gaussian adaptive threshold.

- `max` is the maximum pixel value.

- `Technique` can be either `ADAPTIVE_THRESH_MEAN_C`
       or `ADAPTIVE_THRESH_GAUSSIAN_C`.

- `thresType` is the primary thresholding technique.

- `block` is the size of the block around each pixel.

- `C` is a tupla with constants used for each technique.

Example 10.3 Other binarization methods of an image

As an example consider the image Lena. This image is binarized using the THRESH_BINARY method and is shown in Figure 10.8a. Then in Figure 10.8b the method ADAPTIVE_THRESH_MEAN_C is used, and finally, the method ADAPTIVE_THRESH_GAUSSIAN_C is used and shown in Figure 10.8c. The script is:

```
"' This is script Example_10_3.py
This program binarizes an image.'"

import cv2 as cv
import matplotlib.pyplot as plt

# The image is read:
img = cv.imread('images/figure000.png', 0)

# Script continues at next page.
```

```
# Script continued from previous page.

# THRESH_BINARY is applied:
ret,th1 = cv.threshold(img,80,255,cv.THRESH_BINARY)

# Adaptive threshold mean-C is applied:
th2 = cv.adaptiveThreshold(img,255,\
      cv.ADAPTIVE_THRESH_MEAN_C, v.THRESH_BINARY,11,2)

# Adaptive threshold Gaussian-C is applied:
th3 = cv.adaptiveThreshold(img,255,\
      cv.ADAPTIVE_THRESH_GAUSSIAN_C,cv.THRESH_BINARY,11,2)

# Results are displayed:
cv.imshow('Original Image', img)
cv.imshow('Binary', th1)
cv.imshow('Mean C', th2)
cv.imshow('Gaussian', th3)
cv.waitKey(0)
cv.destroyAllWindows()
```

Example 10.4 Another binarization method of an image
Another method that operates on individual pixels is `bitwise_not`. This method takes a grayscale image and returns the complemented one. For example:

```
"' This is script Example_10_4.py
This program binarizes an image and return the
complemented image.'"
import cv2

# The image is read:
img = cv2.imread('images/breast_digital_Xray.tif',0)

# The image is complemented:
imgComp = cv2.bitwise_not(img)

cv2.imshow('img', img)
cv2.imshow('Bitwise complemented image', imgComp)
cv2.waitKey(0)
cv2.destroyAllWindows()
```

The result for this script `breast_XRay.tif` is shown in Figure 10.9.

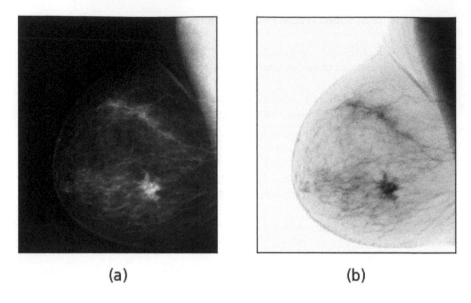

(a)   (b)

FIGURE 10.9: a) Original X-ray image, b) Bit-complemented image.

10.5   Histogram

A histogram is a representation of an image. It can be used to analyze images and objects and video information. Histograms can be used to represent the color distribution of an object, an edge gradient, or a distribution of probabilities to determine the location of an object. Basically, histograms are collected counts of data organized in predefined bins. What is of interest is the number of pixels that fall within a bin range. OpenCV has the method to obtain the histogram. This method is:

```
histogram = cv2.calcHist([image], layer, mask, [Size], [Range])
```

where the parameters are:

- `layer` is the layer number, 0 for grayscale images.

- A `mask` can be used on the image. `None` is used for no mask.

- `Size` is the number of bins in the histogram.

- `Range` is a tuple with the range of values to be measured.

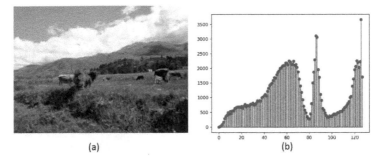

FIGURE 10.10: a) Cows image, b) Histogram.

Example 10.5 Histogram of an image
Consider the image `cows.png` in Figure 10.10a. A script to obtain the histogram that can be plotted with a `stem` plot from `matplotlib.pyplot` is

```
"' This is script Example_10_5.py
This program obtains the histogram of an image.'"
import cv2
from matplotlib import pyplot as plt

img = cv2.imread('images/cows.png', 0)
hist = cv2.calcHist([img], [0], None, [128], (0, 256))
cv2.imshow('img', img)
a = plt.figure(2)
plt.stem(hist)
plt.show()
cv2.waitKey(0)
cv2.destroyAllWindows()
```

The histogram is shown in Figure 10.10b. In the case of color images, the `calcHist` method has to be repeated three times by only changing the layer number to 0, 1, or 2.

10.5.1 Histogram Equalization

In the case of images whose histogram is not evenly distributed, it is possible to perform an equalization with the instruction

```
result = equalizeHist(image)
```

Example 10.6 Histogram equalization
As an example, the image `corn.png` shown in Figure 10.11a, does not have a good contrast. The following script equalizes the image and saves the result in

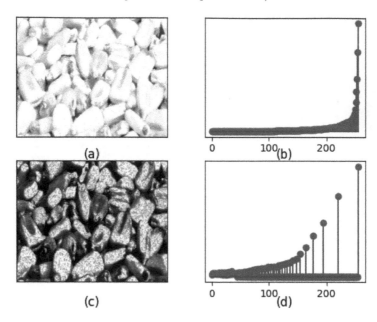

FIGURE 10.11: a) Original corn image, b) original corn histogram, c) equalized corn image, and d) equalized histogram.

`resulting_image`. Its histogram is shown in Figure 10.11b and the resulting image after equalization is shown in Figure 10.11c and the corresponding equalized histogram is shown in Figure 10.11d.

```
"' This is script Example_10_6.py
This program equalizes an histogram.'"

# Import the necessary libraries:
import cv2
from matplotlib import pyplot as plt

# Read the image as a gray level image
# and obtain the histogram:
img = cv2.imread('images/corn_light.jpg', 0)
hist = cv2.calcHist([img], [0], None, [256], (0, 256))

# Script continues at next page.
```

```
# Script continued from previous page.

# Equalize and plot the histogram and the resulting image:
resulting_Image = cv2.equalizeHist(img)
resulting_hist = cv2.calcHist([resulting_Image], [0], None,
[256], (0, 256))
cv2.imshow('img', img)
cv2.imshow('img', resulting_Image)


a = plt.figure(2)
plt.stem(hist)
a = plt.figure(3)
plt.stem(resulting_hist)
plt.show()

cv2.waitKey(0)
cv2.destroyAllWindows()
```

10.6   Draw Geometric Shapes and Text on an Image

It is possible to draw geometric shapes on an image. The shapes that can be drawn are lines, rectangles, circles, ellipses, and polygons. It is also possible to draw text on an image. The methods are:

- cv2.line(image, startPoint, endPoint, color, thickness)

- cv2.circle(image, center, radius, color, thickness)

- cv2.rectangle(image, startPoint, endPoint, color, thickness)

- cv2.ellipse(image, center, axes, angle,startAngle, endAngle, color, thickness, lineType)

- cv2.polylines(image, [point], isClosed, color, thickness)

- cv2.putText(image, text, org, font, fontScale, color, thickness)

The parameters that can be declared in these methods are:

- color: It specifies the color of the shape.

- thickness: It specifies the thickness of the line, circle, etc. The default value is 1. In a closed shape, a value of thickness = -1 fills up the shape.

- lineType: An optional parameter that gives the type of the ellipse boundary.

An example shows how different shapes are drawn on an image.

Example 10.7 Shapes drawn on an image.
In this example, an image is created using numpy. Then, different shapes are drawn on it. First, openCV and numpy are imported

```
import cv2
import numpy as np
```

The image to be created is a white color square created with ones in every pixel and multiplied by 255 to make it a white square as

```
image = np.ones((512, 512, 3), np.uint8)*255
```

Now, a green straight line that starts at the pixel located at (50, 20) and ends at the pixel located at (200, 100), the color has a component green and no components red and blue and the thickness is 3.

```
cv2.line(image, (50, 20), (200, 100), (0, 255, 0), 3)
```

A circle is added with a center at (400, 300), a radius equal to 50 and color components red and blue so the color tuple is (255, 0, 255) and the thickness is 5. The instruction is

```
cv2.circle(image, (400, 300), 30, (255, 0, 255), 5)
```

A rectangle is added with the upper left vertex at (100, 50) and the lower right vertex at (300, 500). The color is blue so the tuple for the color is (255, 0, 0) and the thickness is 2. The instruction is then

```
cv2.rectangle(image, (100, 50), (300, 500), (255, 0, 0), 2)
```

A text "OpenCV" is written on the image with the lower left coordinate at (0, 400). The font is FONT_ITALIC, thickness is equal to 2, and color given by (255, 0, 255) which corresponds to a purple color. The instruction is

```
cv2.putText(image, "OpenCV", (0, 400), cv2.FONT_ITALIC,
            (16, 255, 0, 255), 2)
```

Finally, an ellipse is printed with center at (100, 200), axes size = 70 and 200; thus, the parameter is (70, 200), the startAngle and endAngle are both equal to 0, the color is (55, 100, 100), and the thickness is 2. The ellipse is:

```
cv2.ellipse(image,(100, 200),(70, 200),0, 0, 360,(55, 100, 100), 2)
```

The final script is:

```
''' This is script Example_10_7.py
This script draws text and geometric shapes on an image.'''

# Import the necessary libraries:
import cv2
import numpy as np

# Create the black image:
image = np.zeros((512, 512), 3, np.uint8)
# Create the straight line:
cv2.line(image, (50, 20), (200, 100), (0, 255, 0), 3)

# Create the circle:
cv2.circle(image, (200, 100), 30, (255, 0, 255), 5)

# Create the rectangle:
cv2.rectangle(image, (100, 50), (300, 500) (255, 0, 0), 2)

# Write the text:
cv2.putText(image, "OpenCV", (200,100), 30, (255,0,255), 2)

# Create the ellipse:
cv2.ellipse(image,(100,200),(70,200),0,0,360,(55, 100, 100),2)

cv2.imshow('Drawing', image)
cv2.waitKey(0)
cv2.destroyAllWindows()
```

The image created is shown in Figure 10.12.

10.7 Contour Detection

A task frequently used in image processing is contour detection, also known as shape detection. To do this task, openCV has a method to find the contours and draw them on the image. A required step is to binarize the image. This can be done by first converting the image to a gray-level image and then by binarizing it. This is done first with

```
# The image is converted to a gray level image.
imagGray = cv2.cvtColor(image, cv2.COLOR_BGR2GRAY)
```

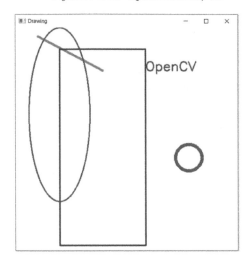

FIGURE 10.12: Image with geometric shapes and text.

```
# The gray image is binarized with the method THRESHOLD
th = cv2.threshold(imagGray, 100, 255, cv2.THRESH_BINARY)
```

The contours can be found with the OpenCV method `findContours` which has the format

```
findContours(image, contour_retrieval, contours_approximation)
```

where the parameters are:

For `contour_retrieval` the parameters are

- `cv2.RETR_TREE` It retrieves all of the contours and hierarchy.

- `cv2.RETR_EXTERNAL` It retrieves outer contours.

- `cv2.RETR_LIST` It retrieves all of the contours without hierarchy.

- `cv2.RETR_CCOMP` It retrieves all of the contours and organizes them in a two-level hierarchy.

For `contours_approximation` the parameters are

- `cv2.CHAIN_APPROX_NONE` All boundary points are saved.

- `cv2.CHAIN_APPROX_SIMPLE` It compresses the contour by saving only the end points.

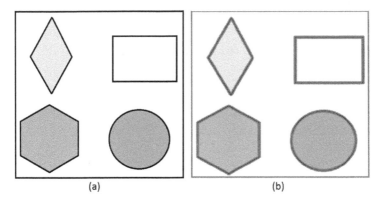

(a)   (b)

FIGURE 10.13: Contours for image with geometric figures, a) original figures and b) contours.

Example 10.8 Contour detection.

Consider the image shown in Figure 10.13a. This image contains several color geometric shapes. A script to find and draw the contours starts importing the necessary libraries, **openCV** and **numpy**,

```
import cv2
import numpy as np
```

The image is read, converted to a gray image, and binarized:

```
image = cv2.imread('figura10_shapes.png')
imgGray = cv2.cvtColor(image, cv2.COLOR_BGR2GRAY)
_, th = cv2.threshold(imgGray, 150,255, cv2.THRESH_BINARY)
```

The contours are found with

```
contours, hierarchy = cv2.findContours(image, cv2.RETR_EXTERNAL,\
    cv2.CHAIN_APPROX_NONE)
```

The contours are plotted with the method **drawContours(image, contours, contourldx, color, thickness)** on the original image. The parameters are:

- contours: Set of contours found with **findContours**.

- contourldx: Indicates a contour to be drawn. A negative value draws every contour.

- color: Color of the contours.

- thickness: Thickness of the drawn contour.

Thus, the method `drawContours` is

```
drawContours(image, contours, -1, (0, 255, 0), 3)
```

Finally, the image is displayed

```
cv2.imshow('Original', image)
cv2.imshow('Contours', img)
cv2.waitKey(0)
cv2.destroyAllWindows()
```

The complete script is

```
''' This is script Example_10_8.py
This program obtains and plots the contours.'''

import cv2
import numpy as np

# The image is read, converted to a gray image,
# and binarized:

image = cv2.imread('figura10_shapes.png')
imgGray = cv2.cvtColor(image, cv2.COLOR_BGR2GRAY)
_, th = cv2.threshold(imgGray, 150,255, cv2.THRESH_BINARY)

# The contours are found with

contours, hierarchy = findContours(image, cv2.RETR_EXTERNAL, \
        cv2.CHAIN_APPROX_NONE)

# The contours are plotted with drawContours

drawContours(image, contours, -1, (0, 255, 0), 3)

# Finally, the image is shown

cv2.imshow('Example', image)
cv2.imshow('th', th)
cv2.imshow('img', img)
cv2.waitKey(0)
cv2.destroyAllWindows()
```

The contours are shown in Figure 10.13b.

10.8 Frequency Domain Processing

The basic operator in the frequency domain is the Fourier transform which in digital image processing is called the 2-D Discrete Fourier Transform (DFT). For an $N \times M$ image represented by a function $f(x, y)$ and for $x = 0, 1, 2, \ldots, M - 1$ and $y = 0, 1, 2, \ldots, N - 1$, the DFT is defined by the equation

$$F(u, v) = \sum_{x=0}^{M-1} \sum_{y=0}^{N-1} f(x, y) e^{-j2\pi(ux/M + vy/N)} \tag{10.1}$$

for $u = 0, 1, 2, \ldots, M - 1$ and $v = 0, 1, 2, \ldots, N - 1$. The frequency domain is the coordinate system spanned by $F(u, v)$ with u and v being the frequency variables. The inverse discrete Fourier transform is given by

$$f(x, y) = \frac{1}{MN} \sum_{u=0}^{M-1} \sum_{v=0}^{N-1} F(u, v) e^{j2\pi(ux/M + vy/N)} \tag{10.2}$$

for $x = 0, 1, 2, \ldots, M - 1$ and $y = 0, 1, 2, \ldots, N - 1$.

Evaluation of the DFT requires a number of multiplications in the order of $M^2 N^2$. Since multiplications are time consuming, an algorithm was developed to optimize the number of multiplications. This is made with an algorithm that is called the Fast Fourier transform (FFT) which reduces the number of multiplications to the order of $MN log_2(MN)$. OpenCV and numpy have methods dft and fft, respectively, to calculate the FFT.

Example 10.9 Discrete Fourier transform
Consider the image shown in Figure 10.14a which is a simple black rectangle in a white background. In order to compute the DFT, the image has to be converted to a floating point image using the numpy method np.float32(img). Then, the DFT is computed with dft and the result is a complex number. To center the spectrum of the DFT a shift must be made, numpy has a method np.fft.fftshift(dft) to shift the DFT. Finally, the image and the magnitude of the DFT can be plotted with

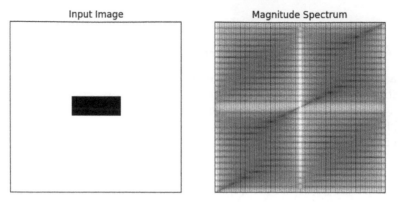

FIGURE 10.14: a) Original figure, b) The discrete Fourier transform.

```
'' This is script Example_10_9.py
This script obtains the Fourier transform of an image.'"
import numpy as np
import cv2
from matplotlib import pyplot as plt

img = cv2.imread('images/cuadro.png', 0)
img_float32 = np.float32(img)

dft = cv2.dft(img_float32, flags = cv2.DFT_COMPLEX_OUTPUT)
dft_shift = np.fft.fftshift(dft) + 1e-9
magn = 20*np.log(cv2.magnitude(dft_shift[:,:,0],dft[:,:,1]))

plt.subplot(121)
plt.imshow(img, cmap = 'gray')
plt.title('Input Image')
plt.xticks([])
plt.yticks([])
plt.subplot(122)
plt.imshow(magn, cmap = 'gray')
plt.title('Magnitude Spectrum')
plt.xticks([])
plt.yticks([])
```

The image and the magnitude of the DFT are shown in Figure 10.14. To evaluate the inverse DFT (IDFT), first shift again to restore the DFT and then compute the IDFT as

```
f_ishift = np.fft.ifftshift(dft_shift)
img_back = cv2.idft(f_ishift)
```

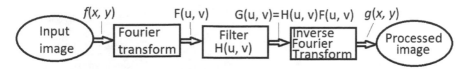

FIGURE 10.15: Filtering process with two-dimensional FFT.

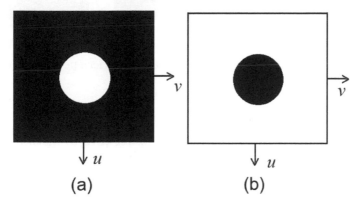

FIGURE 10.16: a) Low-pass filter, b) High-pass filter.

As can be seen from Figure 10.15 there are two regions of interest, the white and the black regions are regions where there are no changes in frequency and they correspond to low frequencies, however, there are sharp changes at the edges of the black rectangle which correspond to high frequencies.

Example 10.10 Filtering in the frequency domain
Filtering is used to remove unwanted signals such as noise or frequency components that are not of interest in the processing. There are several types of filters and they are low-pass, high-pass, bandpass, and band-reject. Low-pass filters remove low-frequency signals such as those inside and outside the square as Figure 10.14a. High-pass filters remove high-frequency signal such as the ones in sudden changes in gray level as those at the edge of the black square. Bandpass filters only remove a certain band of frequencies, and band-reject filters remove a band of unwanted signals. Figure 10.16 shows ideal low-pass and high-pass filters, the black region corresponds to the rejected frequencies.

The filtering process is carried upon the image in the following way. First the required libraries are imported,

```
import cv2
import numpy as np
from matplotlib import pyplot as plt
```

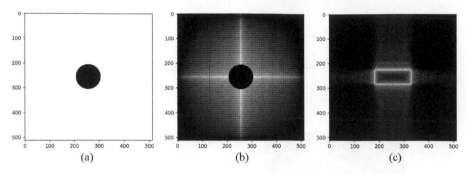

FIGURE 10.17: a) High-pass filter, b) Filtered Fourier spectrum, c) Filtered image.

The image to filter is read in and resized to 512×512 to make it a square image and the size a power of 2, in this case 2^8:

```
image = cv2.imread("square.png")
Nu = 512
Nv = 512
image = cv2.resize(image, (Nu, Nv))
```

The image has to be converted to a gray-level image and converted to a floating-point image

```
imagGray = cv2.cvtColor(image, cv2.COLOR_RGB2GRAY)
imagGrayF = np.float64(imagGray)
```

The Fourier transform is taken on the floating-point image `imagGrayF` and the transform is shifted to the center of the spectrum:

```
original = np.fft.fft2(imagGrayF)
center = np.fft.fftshift(original)
```

The log scale of the absolute value of the Fourier transform is computed and displayed:

```
plt.subplot(152)
plt.imshow(np.log(1 + np.abs(center)), "gray")
plt.title("Spectrum")
```

The spectrum is shown in Figure 10.17a. The design of the high-pass filter is done with an ideal filter. This filter passes the high frequencies, those frequencies far from the origin, and blocks or rejects the low-frequency signals close to the center of the frequency spectrum. The design of the filter requires a matrix of ones (1's) with the size of the image. The number of rows and columns is computed and the center is established. With a `for` loop, the points inside the filter reject band are made 0 (zero). The filter function is in the variable `base`. These operations are performed in a method or function as:

```
def idealFilterHP(D0, imgShape):
    base = np.ones(imgShape[:2])
    rows, cols = imgShape[:2]
    center = (rows/2, cols/2)
    for x in range(cols):
        for y in range(rows):
            if distance((y, x), center) < D0:
                base[y, x] = 0
    return base
```

The filter method requires a calculation of the distance in a method `distance` whose arguments are the points. This method is:

```
def distance(p1, p2):
    return np.sqrt((p1[0]-p2[0])**2 + (p1[1]-p2[1])**2)
```

The method `idealFilterHP` is called with a radius of 50 and the frequency response magnitude is plotted as

```
HighPass = idealFilterHP(50, img.shape)
plt.subplot(153)
plt.imshow(np.log(1+np.abs(HighPass)), "gray")
plt.title("High Pass Filter")
```

As can be seen in Figure 10.15, the filtering process is a multiplication of Fourier transforms. This is realized as

```
filteredSignal = center*HighPass
```

The `filteredSignal` variable is the DFT of the result which is plotted with:

```
plt.subplot(154)
plt.imshow(np.log(1 + np.abs(filteredSignal)),"gray")
plt.title("Spectrum of result")
```

The filtered signal is shown in Figure 10.17b. The figure shows the absence of low-frequency components. Finally, the inverse Fourier transform is computed and the filtered image is displayed as Figure 10.17c.

```
filteredSignal = np.fft.ifftshift(filteredSignal)
inverseFilteredSignal = np.fft.ifft2(filteredSignal)
plt.subplot(155)
plt.imshow(np.log(1 + np.abs(inverseFilteredSignal)), "gray")
plt.title("Processed image")
plt.show()
```

Figure 10.17c shows the resulting filtered image. It can be seen that only the region of high-frequency content, namely the edges, is preserved. The complete script is

```
"' This is script Example_10_10.py
This script filters an image in the frequency domain.'"

# Import the libraries needed:
import cv2
import numpy as np
from matplotlib import pyplot as plt

# Method for the filter definition:
def idealFilterHP(D0, imgShape):
   base = np.ones(imgShape[:2])
   rows, cols = imgShape[:2]
   center = (rows/2, cols/2)
   for x in range(cols):
      for y in range(rows):
         if distance((y, x), center) < D0:
            base[y, x] = 0

   return base

# The method distance whose arguments are the points p1 and 2:
def distance(p1, p2):
   return np.sqrt((p1[0]-p2[0])**2 + (p1[1]-p2[1])**2)

# The input image is read in and resized:
image = cv2.imread("square.png")
Nu = 512
Nv = 512
image = cv2.resize(image, (Nv, Nc))

# Script continues at next page.
```

```
# Script continued from previous page.

# The image is converted to gray level and floating-point:
imagGray = cv2.cvtColor(image, cv2.COLOR_RGB2GRAY)
imagGrayF = np.float64(imagGray)

# Fourier transform computation and shifting:

original = np.fft.fft2(imagGrayF)
center = np.fft.fftshift(original)

# Plotting of the Fourier spectrum for the input image:
plt.subplot(152)
plt.imshow(np.log(1 + np.abs(center)), "gray")
plt.title("Spectrum")

# The high pass filter is designed with D0 = 50 and plotted:
HighPass = idealFilterHP(50, img.shape)
plt.subplot(153)
plt.imshow(np.log(1+np.abs(HighPass)), "gray")
plt.title("High Pass Filter")

# The filtering process is a multiplication
# of Fourier transforms:
filtered_Signal = center*HighPass

# The filtered_Signal is the DFT of the result. Its plot is:

plt.subplot(154)
plt.imshow(np.log(1 + np.abs(filtered_Signal)), "gray")
plt.title("Spectrum of result")

# The inverse Fourier transform is computed.
# The filtered image is plotted:
filtered_Signal = np.fft.ifftshift(filtered_Signal)
inverseFilteredSignal = np.fft.ifft2(filtered_Signal)
plt.subplot(155)
plt.imshow(np.log(1 + np.abs(inverseFilteredSignal)), "gray")
plt.title("Processed image")
plt.show()
```

The filtered signal shows that only in the region where there is abrupt change, that corresponds to high frequencies, is kept. In the regions where there is no change; that is, where there are only black or only white, that correspond to lower frequencies, are eliminated. The oscillations that appear in the response are known as Gibbs effect and are due to the ideal filter. They are eliminated if a non-ideal type of filter, such as the Butterworth filter, is used. The Butterworth low-pass filter has a smooth transition from the pass band to the reject band and its magnitude is given by

$$T(u, v) = \frac{1}{1 + D(u, v)/D0]^{2n}} \tag{10.3}$$

where $D(u, v)$ is the distance from the origin which is implemented with the method `distance` from above. A method that implements the low-pass Butterworth function is:

```
def ButterworthLP(D0, imgShape, n):
    filter = np.zeros(imgShape[:2])
    rows, cols = imgShape[:2]
    center = (rows/2, cols/2)
    for x in range(cols):
        for y in range(rows):
            filter[y, x] = 1/(1 + (distance((y, x), center)/D0)**(2*n))
    return filter
```

The high-pass filter can be implemented changing the second row to

```
filter = np.ones(imgShape[:2])
```

The high pass method is then

```
def ButterworthHP(D0, imgShape, n):
    filter = np.ones(imgShape[:2])
    rows, cols = imgShape[:2]
    center = (rows/2, cols/2)
    for x in range(cols):
        for y in range(rows):
            filter[y, x] = 1/(1 + (distance((y, x), center)/D0)**(2*n))
    return filter
```

The other change that has to be made is

```
HighPass = ButterworthHP(50, img.shape)
```

The result of the filtering with these changes produces Figure 10.18. It can be seen that the Butterworth filter reduces the Gibbs effect.

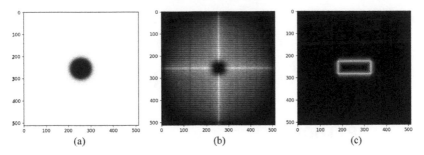

FIGURE 10.18: a) High-pass filter, b) Filtered Fourier spectrum, c) Filtered image.

10.9 Noise Addition to Images

To study noise removal in images it is useful to have a way to add noise to them. There is a method in `scikit-image` that can do this. This method is `random_noise` and it can be called with:

```
from skimage.util import random_noise
random_noise(image, mode = 'Choice',seed = None, clip= True)
```

where `Choice` can be one mode from the list:

1. `gaussian` - Gaussian-distributed additive noise.

2. `localvar` - Gaussian-distributed additive noise with specified local variance.

3. `poisson` - Poisson-distributed noise generated from data.

4. `salt` - Replaces random pixels with 1.

5. `pepper` - Replaces random pixels with 0.

6. `s&p` - Replaces random pixels with 0 or 1.

7. `speckle` - Multiplicative noise with specified mean and variance.

The resulting noisy image is a floating-point image. If the input image is a `uint8` grayscale image, the resulting image can be converted back to the original format with `skimage.imgas_ubyte`. The parameter `seed` is used in case it is desired to initiate the random process with a seed. The parameter `clip` is a boolean. If it is `True` (default) the output is clipped to keep the pixels between 0 and 255.

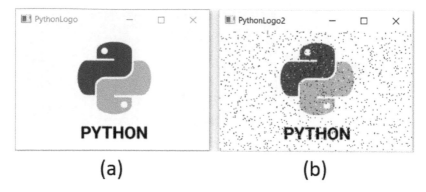

FIGURE 10.19: a) Original figure, b) Salt and pepper noise added.

Example 10.11 Noise in images

As an example consider the image shown in Figure 10.19a where salt and pepper noise is added. The instructions are:

```
"' This is script Example_10_11.py
This script adds noise to an image.'"

import cv2
from skimage.util import random_noise

Logo = cv2.imread("PythonLogo.png", 0 )
cv2.imshow("PythonLogo", Logo)
image = random_noise(Logo, mode = 's&p',seed=None, clip=True)
cv2.imshow("PythonLogo2", image)
cv2.waitKey(0)
cv2.destroyAllWindows()
```

The resulting noisy image is shown in Figure 10.19b.

10.9.1   Denoising

There are several methods for denoising, the attempt to remove noise from an image. The methods that are used in this section are: `filter2D`, `blur`, and `gaussianBlur`. The method filter2D has the format

```
image = cv2.filter2D(source, depth, kernel)
```

where source is the noisy image, **depth** is the type of the image, for example, int, uint8, float, etc. Use -1 to have the same type in the processed image, and the kernel is a matrix for the filtering that may be a 3×3, 5×5, 7×7, etc, but always an odd number of rows and columns.

Example 10.12 Noise removal of an image

In this example the image **Lena** is used. As always the necessary libraries are imported.

```
import cv2
from skimage.util import random_noise
import numpy as np
```

Then the image **Lena** is imported and salt and pepper noise is added. The noisy image is displayed in Figure 10.20

```
Lena = cv2.imread('lenaC.jpg', 0)
cv2.imshow('lena', Lena)
image = random_noise(Lena, mode = 's&p',seed = None, clip = True)
cv2.imshow('Noisy image', image)
```

The kernel is defined and the filter2D is applied. The filtered image is displayed and shown in Figure 10.20b.

```
kernel = np.ones((3,3), np.float32)/9
filt2d = cv2.filter2D(image, -1, kernel)
cv2.imshow('Filtered', filt2d)
```

The next method that is applied is the blur method. This is implemented with

```
blur = cv2.blur(image, (3, 3))
cv2.imshow('Blur', blur)
```

which is displayed in Figure 10.20c. The last method is the Gaussian blur implemented and displayed in Figure 10.20d with

```
gaussian_blur = cv2.GaussianBlur(image, (5,5), 0)
cv2.imshow('Gaussian blur', gaussian_blur)
```

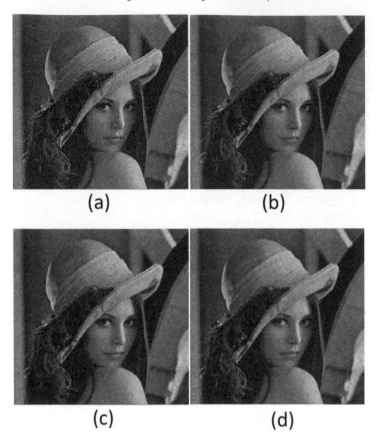

FIGURE 10.20: a) Image with salt and pepper noise, b) Image filtered with filter2D, c) Image blurred, d) Image processed with Gaussian blur.

The final script is

```
"' This is script Example_10_12.py
This script filters an image to remove noise.'"

import cv2
from skimage.util import random_noise
import numpy as np

# The image Lena is read and salt and pepper noise is added.

Lena = cv2.imread('lenaC.jpg', 0)
cv2.imshow('lena', Lena)
image = random_noise(Lena, mode = 's&p',seed=None,clip=True)
cv2.imshow('Noisy image', image)

# Script continues at next page.
```

```
# Script continued from previous page.

# The kernel is defined and the filter2D is applied:
kernel = np.ones((3,3), np.float32)/9
filt2d = cv2.filter2D(image, -1, kernel)
cv2.imshow('Filtered', filt2d)

# The blur method:
blur = cv2.blur(image, (3, 3))
cv2.imshow('Blur', blur)

# The last method is the Gaussian blur:
gaussian_blur = cv2.GaussianBlur(image, (5,5), 0)
cv2.imshow('Gaussian blur', gaussian_blur)

cv2.waitKey(0)
cv2.destroyAllWindows()
```

10.10 Morphological Image Processing

Morphological image processing is a set of techniques with the purpose of analyzing and extracting the geometric properties of an image. The mathematical tool for this purpose is mathematical morphology extracting image components [1]. In this way it is possible to obtain a coherent mathematical description of the geometrical structures in the image. Set theory is the language of mathematical morphology. It is assumed the reader has a basic knowledge of set theory and the additional definitions are presented here. In this section we define and implement operations such as erosion, dilation, opening, closing, thinning, and skeletonization. These operations require the use of a kernel or structuring element that must be defined prior to processing.

10.10.1 Erosion and Dilation

The concepts of reflection and translation are given now. The reflection a set B is denoted by $\hat{B}$ is defined as

$$\hat{B} = \{x|x = -a \text{ for } a \in B\} \tag{10.4}$$

The reflection set $\hat{B}$ is then the set of points in B whose coordinates (x, y) are replaced by $(-x, -y)$. The translation of a set B by a point $z = (x, y)$ is

1	1	1
1	1	1
1	1	1

0	1	0
1	1	1
0	1	0

0	1	1	0
1	1	1	1
0	1	1	0

     (a)   (b)   (c)

FIGURE 10.21: Structuring elements. a) Square, b) Circular, c) Elliptical.

defined as

$$B_z = \{w|w = b + z \text{ for } a \in B\} \tag{10.5}$$

A structuring element is a small set or subimage used to process the image of interest. Some structuring elements are shown in Figure 10.21.

10.10.2   Dilation and Erosion

The erosion of a set A by a structuring element B is defined by

$$A \ominus B = \{z|B_z \subseteq A\} \tag{10.6}$$

This means that for every point in the image, the reflection of the structuring element fits completely in the resulting image. As an example consider the image and structuring element in Figure 10.22. The result is an image that looks smaller; that is, it has been eroded. As an example, the erosion process is shown in Figure 10.23. In Figures 10.23a and c the structuring element does not fit completely in the image and thus those pixels are eroded from the image as shown in Figures 10.23b and d. In Figure 10.24a, b, and c, the structuring element does fit inside the image and these pixels are preserved in the erosioned image which is shown in Figure 10.24d.

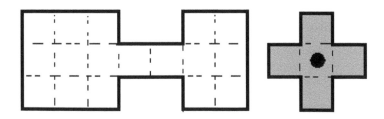

FIGURE 10.22: Image and structuring element.

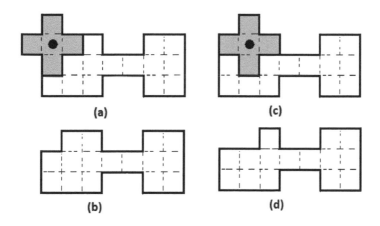

FIGURE 10.23: Erosion process initiated.

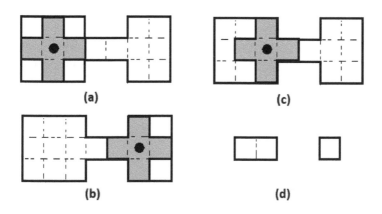

FIGURE 10.24: Erosion process terminated.

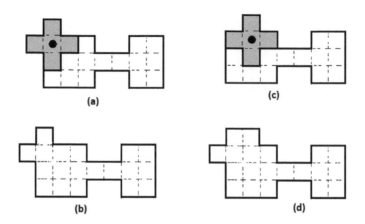

FIGURE 10.25: Dilation process.

The dilation of a set A by an structuring element B is defined by

$$A \oplus B = \{z | \hat{B}_z \cap A \neq \varnothing\} \tag{10.7}$$

In this case the structuring element does not have to fit inside the original image. The result is an image that dilates according to the structuring element. Figure 10.25a shows a pixel where the structuring element dilates the image as shown in Figure 10.25b. Figures 10.25c and d show another pixel and the resulting dilated image is shown in Figure 10.26.

The morphological operations are best used with gray level images and then a method for edge detection is used for demonstration. An example shows the technique.

Example 10.13 Dilation and erosion of an image
The dilation and erosion is done on the Lena image. First the necessary libraries are imported:

```
import cv2
import numpy as np
```

Read the image Lena as a gray-level image

```
Lena = cv2.imread('lenaC.jpg', 0) # Read as a gray level image.
```

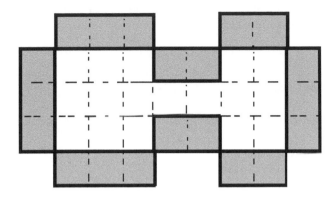

FIGURE 10.26: Dilation process.

The edges of the image are obtained with an alternative method known as the Canny detector named after his author [1]. The Canny method requires the source image and two threshold values, minimum and maximum, which in this example are chosen as 150 and 200. The processed image is then displayed:

```
# Edgedetection using Canny detection
LenaCanny = cv2.Canny(Lena, 150, 200)
cv2.imshow('Lena edges', LenaCanny)
```

The dilation is implemented with the OpenCV method dilate whose format is

```
dilate(source, kernel, No. of iterations)
```

The source image is LennaCanny, the kernel is a square kernel of size 3; that is, a matrix of ones of dimension 3×3 with integer type, and the number of iterations is 1.

```
# Dilation
kernel = np.ones((3,3), np.uint8)
LenaDilation =cv2.dilate(LenaCanny, kernel,iterations = 1)
cv2.imshow('Lena dilated', LenaDilation)
```

The erosion is implemented with the method `erode` which has the same parameters as the method `dilate`

```
# Erosion
LenaErosion = cv2.erode(LenaDilation, kernel,iterations = 1)
cv2.imshow('Lena eroded', LenaErosion)
```

Finally, the last two instructions are to display and close the images,

```
cv2.waitKey(0)
cv2.destroyAllWindows()
```

The complete script is:

```
"' This is script Example_10_13.py
This script dilates and erodes an image.'"

# Import the necessary libraries:
import cv2
import numpy as np

# Read the image Lena as a gray level image:
Lena = cv2.imread('lenaC.jpg', 0) # Read as a gray level image.

# Edgedetection using Canny detection
LenaCanny = cv2.Canny(Lena, 150, 200)
cv2.imshow('Lena edges', LenaCanny)

# Dilation
kernel = np.ones((3,3), np.uint8)
LenaDilation =cv2.dilate(LenaCanny, kernel,iterations = 1)
cv2.imshow('Lena dilated', LenaDilation)

# The erosion is implemented with the method erode:
# Erosion
LenaErosion = cv2.erode(LenaDilation, kernel,iterations = 1)
cv2.imshow('Lena eroded', LenaErosion)

# The last two instructions are to display and
# close the images
cv2.waitKey(0)
cv2.destroyAllWindows()
```

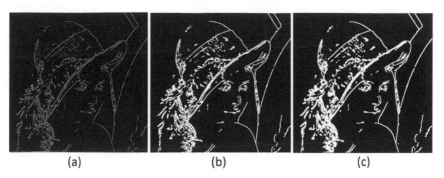

(a) (b) (c)

FIGURE 10.27: a) Edge detection with a Canny detector, b) Dilation of the image, c) Erosion of the dilated image.

The results for the Canny edge detection, the dilation, and the erosion are shown in Figure 10.27.

10.11 Python Instructions in Chapter 10

Table 10.2 shows the instructions presented in Chapter 10.

TABLE 10.2: Python instructions for Chapter 10

Instruction	Description
imread	It reads an image from a file.
waitKey	It indicates for how long the image is displayed.
destroyAllWindows	It closes the images displayed.
shape	It gives the shape of an image.
cvtColor	It converts an image to another format.
capture	It capture video frames.
VideoCapture	It captures video.
VideoWriter	It writes video to a file.
threshold	It indicates the threshold for binarization.
adaptiveThreshold	It defines a threshold function.
bitwise_not	It is used for binary complement an image.
calcHist	It obtains the image histogram.
stem	It implements a stem plot.
equalizeHist	It equalizes an histogram.
line	It draws a line on an image.
circle	It draws a circle on an image.
rectangle	It draws a rectangle on an image.
ellipse	It draws an ellipse on an image.
polylines	It draws polylines on an image.

(Continues next page.)

TABLE 10.2:   Python instructions in Chapter 10 (*Continued*)

Instruction	Description
putText	It writes text on an image.
putText	It writes text on an image.
findContours	It finds contours.
drawContours	It draws contours on an image.
dft	Discrete Fourier transform.
fft	Discrete Fourier transform.
fftshift	It centers the origin of the Fourier transform.
resize	It changes the size of an image.
High-pass filter	It filters an image removing low frequencies.
Low-pass filter	It filters an image removing high frequencies.
Salt and pepper	Noise that produces white and black pixels.
filter2D	It provides filtering in the frequency domain.
GaussianBlur	It provides a blur using a Gaussian distribution.
erosion	A morphological method to erode an image.
dilation	A morphological method to dilate an image.

10.12   Conclusions

OpenCV provides a great deal of methods and functions to perform operations on images and videos. In this chapter only a few of those methods have been presented. It is the author's expectation that it will motivate readers to explore and use OpenCV for image and video processing.

10.13   Selected Bibliography

[1] Canny, J., A Computational Approach To Edge Detection, IEEE Transactions on Pattern Analysis and Machine Intelligence, Vol. 8, No. 6, pp. 679–698, 1986.

Chapter 11

Machine Learning

This chapter gives an introduction to Machine Learning using Python. A definition of machine learning was given in 1959 by Arthur Samuel. He defined machine learning as "the field of study that gives computers the ability to learn without being explicitly programmed".

Professor Tom Mitchell from the Machine Learning Department at Carnegie Melon University, in 1997, defined machine learning by saying that "a computer program is said to learn from experience E with respect to some task T and some performance measure P, if its performance on T, as measured by P, improves with experience E".

Several examples of machine learning exist in everyday life. For example, the spam filter in e-mail servers, recommendation systems that display ads when surfing the web, the detection of transactions that may look fraudulent, the prediction of revenues in future months, to name a few. The purpose of the chapter is to show the use of Python in machine learning.

11.1   Types of Machine Learning Systems

There are several different types of learning algorithms. The two main types are supervised and unsupervised learning. These two types of learning have also subtypes that allow a better and more refined learning process.

11.1.1   Supervised Machine Learning

The term *Supervised Machine Learning*, or simply *supervised learning*, refers to the fact that an algorithm is fed with a data set in which "right answers" are given. As an example of supervised learning consider the prices of houses given with respect to size. Table 11.1 shows the size and price of a set of houses. Figure 11.1 shows a plot of price vs. size for a set of houses in Tucson, AZ. The price is denoted by the variable y and the house size is the variable x. The values of the set of prices x, called the *features*, are denoted by $x^{(i)}$ and the values of the set y are called the *targets* and are denoted by $y^{(i)}$.

The question to be answered is: What is the price of a house with a size that is not in the table? The answer is given by *machine learning*. The task is

DOI: 10.1201/9781003222118-11

TABLE 11.1:   Data for houses
on sale in Tucson, AZ.

Size (sq. ft.)	Price (x $1000)
1010	260
976	249.9
1538	335
1501	220
1188	215
2010	350
1402	190
1445	288
957	175
1503	225
1010	259.9
1368	279
2198	553
2051	483
1395	320

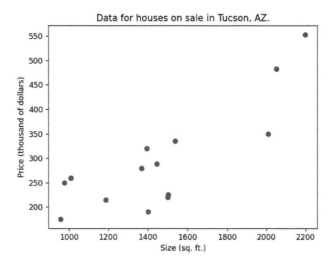

FIGURE 11.1: Tucson house prices vs size in sq. ft.

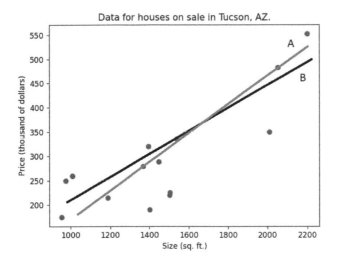

FIGURE 11.2: Training data plot with two different functions.

to search for a function that can predict the price of such a house. A possibility, among many others, might be a straight line. The equation of a straight line is

$$y(x, b, w) = b + wx \qquad (11.1)$$

where b and w are called the parameters of $y(x, b, w)$. The parameter b corresponds to the y-intercept and the parameter w is the slope of the straight line. The straight-line function then approximates the values of

$$\begin{aligned} \hat{y}^{(i)} &= f(x^{(i)}, b, w) \\ &= b + wx^{(i)} \end{aligned} \qquad (11.2)$$

where $\hat{y}^{(i)}$ denotes the predicted values. The set of values $(x^{(i)}, y^{(i)})$ is called the training set. It is assumed that there are m training sets. The example just described is known as a **regression** problem. The term regression refers to the fact that a real-valued output, namely the price, is being predicted. More formally, in supervised learning, a data set, called a *training set*, is used to learn how to predict house prices. Then, the features are used to predict the house prices $\hat{y}^{(i)}$. Figure 11.2 presents the training data together with two possible straight lines corresponding to two hypothesis functions produced by two sets of parameters $\{b$ and $w\}$. The question is: What is the best line to fit the training data? To answer this question, a **cost function** to measure the error has to be defined. Usually, a squared difference is used in the definition of the cost function. The idea is to sum up the squared differences between the real target value and the predicted by the function in equation 11.2, sum up the errors and minimize it. As an example, consider the training set consisting

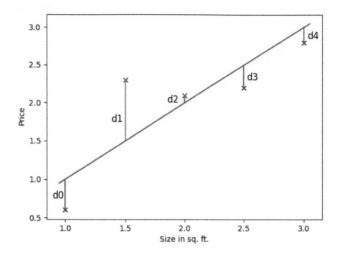

FIGURE 11.3: Errors for a five-point training data. The straight line is the function given by Eq. 11.1.

of five features and five targets shown in Figure 11.3. The squared error sum is given by

$$\sum_{i=0}^{4} d_i^2 = \sum_{i=0}^{4} [f(x^{(i)}, b, w) - y^{(i)}]^2 \tag{11.3}$$

For simplicity, $f(x^{(i)}, b, w)$ is changed to $f_{b,w}(x^{(i)})$. The squared error sum is called the *cost function* and is defined in the following equation as:

$$J(x^{(i)}, b, w) = \frac{1}{2M} \sum_{i=0}^{M-1} [f_{b,w}(x^{(i)}) - y^{(i)}]^2 \tag{11.4}$$

where M is the number of training points. The function $J(x^{(i)}, b, w)$ has to be minimized. Then the task is to find the parameters b and w such that the cost function is minimized. To have an idea of what is the shape of a typical cost function, a plot of it for the data in Figure 11.3 is shown in Figure 11.4. The task is then to minimize the cost function using a suitable optimization algorithm, for example any one of those algorithms in Chapter 9. The optimization process looks for the better straight line that minimizes the cost function.

The model for supervised learning is shown in Figure 11.5. This process is known as the *learning process*. In this figure the training set is fed to the learning algorithm that will find the optimum values for the parameters b and w and use the function given by equation 11.2. The optimization algorithm determines which one is the best set of parameters and predicts a value for the target variable y. In this case the optimization process has obtained a minimum value for $J(x^i, b, w)$.

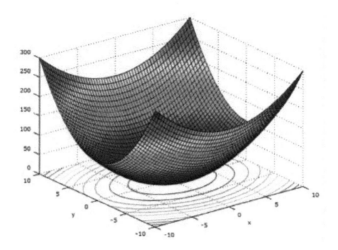

FIGURE 11.4: Cost function plot.

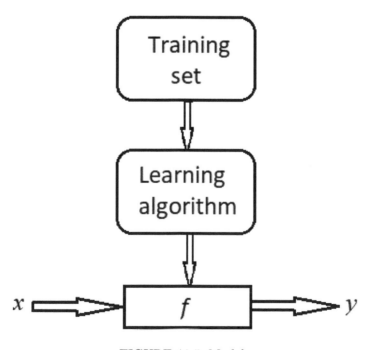

FIGURE 11.5: Model.

TABLE 11.2: Data for houses on sale

Training Example	Size (ft²)	Number of bedrooms	Number of bathrooms	Age of house in years	Price ($1000)
	x_0	x_1	x_2	x_3	y
0	2000	3	4	12	200
1	1527	4	4	15	150
2	1858	3	4	14	250
$\vdots$	$\vdots$	$\vdots$	$\vdots$	$\vdots$	$\vdots$
m	987	2	2	17	89

This approach for predicting a target value from a training set is called a **regression** problem. In a regression problem the goal is to predict a continuous-valued output. In the example above a single feature was considered, namely the size of the house. Thus, this is called a **Univariate** linear regression problem.

11.1.2   Multiple Features

In most cases there are more features in the description of the problem. For example, in the case of house pricing some other features could be the number of bedrooms, the number of bathrooms, the number of floors, the age of the house, and so on. The notation to be used is the following: each feature i for a training example j is denoted by $x_j^{(i)}$ and

n = number of features.
$x^{(i)}$ = set of features of the ith training example.
$x_j^{(i)}$ = value of feature j of the ith training example.

The data for the house could be arranged as shown in Table 11.2. In this table it can be seen that there are $m+1$ training examples. Then the value for $x_j^{(i)}$ for the third feature of the second training example is $x_3^{(2)} = 14$, and the data for the third training example is $x^{(2)} = [2, 1858, 3, 4, 14, 250]$.

The form of the function is now

$$f_w(x) = w_0 x_0 + w_1 x_1 + w_2 x_2 + w_3 x_3 \tag{11.5}$$

For the general case of n features the function $f_{b,w}(x)$ can be written as

$$f_{b,w}(x) = w_0 x_0 + w_1 x_1 + \cdots + w_{n-1} x_{n-1} \tag{11.6}$$

For convenience, the value of x_0 is defined as $x_0 = 1$. Equation 11.6 can be written as

$$f_{b,w}(x) = W^T X \tag{11.7}$$

TABLE 11.3: Data for houses on sale

Training Example	Size ×1000 ft^2	Number of bedrooms	Number of bathrooms	Age in years	Price ($100,000)
	x_0	x_1	x_2	x_3	y
0	2	3	4	12	2
1	1.527	4	4	15	1.50
2	1.858	3	4	14	2.50
⋮	⋮	⋮	⋮	⋮	⋮
m	0.987	2	2	17	0.89

where

$$W = \begin{bmatrix} w_0 \\ w_1 \\ w_2 \\ \vdots \\ w_{n-1} \end{bmatrix}, \quad X = \begin{bmatrix} 1 \\ x_1 \\ x_2 \\ \vdots \\ x_{n-1} \end{bmatrix} \tag{11.8}$$

Both of these vectors are defined in $\mathbb{R}^n$.

This problem where there are several features is called **Multivariate** Linear Regression. The cost function for this case is given by

$$J(b, W) = \frac{1}{2m} \sum_{i=0}^{m} [y^{(i)} - f_{b,w}(x^{(i)})]^2 \tag{11.9}$$

As in the single feature case, the cost function can be minimized using, for example, any of the optimization techniques from Chapter 9. For the case of linear regression, a very popular minimization algorithm is the *gradient descent* algorithm which is described in the next section.

11.1.3 Feature Scaling

Referring to the features shown in Table 11.2, it is noted that the size of the house is in the thousands, the price is in the hundreds, and all other feature values are very small. These numbers can be troublesome for most optimization algorithms. It is recommended to scale the features to be in the same order of magnitude. For example, divide the size by 1000 in this feature, and do the same for all the other features. A better set of values is shown in Table 11.3. Most algorithms would behave better with these values which are obtained after feature scaling.

11.2   Gradient Descent Algorithm

The Gradient descent optimization algorithm is very popular to optimize regression problems. As it was mentioned before, it is needed to minimize the cost function $J(b, w)$ which is expressed as

$$\min_{b,w} J(b, w) \qquad\qquad (11.10)$$

The minimization process goes this way:

1. Start with $J(b, w)$ for a set of initial values for $\{b, w\}$.

2. Change $\{b, w\}$, and keep changing them, to minimize $J(b, w)$.

3. Use the gradient descent algorithm to minimize $J(b, w)$. In this algorithm the cost function is minimized by evaluating

$$w_{j,new} = w_j - \alpha \frac{\partial J(b, w)}{\partial w_j} \qquad\qquad (11.11a)$$

$$b_{new} = b - \alpha \frac{\partial J(b, w)}{\partial b} \qquad\qquad (11.11b)$$

$$\qquad\qquad (11.11c)$$

4. Simultaneously update w and b as

$$w_j = w_{j,new} \qquad\qquad (11.12a)$$

$$b = b_{new} \qquad\qquad (11.12b)$$

5. Continue repeating steps 2 and 3 until $J(b, w)$ reaches a minimum. A minimum is reached when the difference between consecutive values of b, w is less than a small value ϵ, say $\epsilon < 1 \times 10^{-06}$. Thus, the minimization is reached when

$$|b_{new} - b| < \epsilon \qquad\qquad (11.13a)$$

$$|w_{j,new} - w| < \epsilon \qquad\qquad (11.13b)$$

In equations 11.11 the parameter α is called the *learning rate*, it is a positive number and its value is chosen to improve convergence of the technique.

Caution must be observed when choosing the value of α, a very small value for it results in a slow convergence, whereas a large value can lead to non convergence of the optimization algorithm. It might be possible to have an α-value for b and a different α-value for w to improve the convergence of the algorithm.

For linear regression the cost function is given by equation 11.4 and the hypothesis by equation 11.2. Thus, the cost function is given by

$$J(b, w) = \frac{1}{2m} \sum_{i=0}^{m} [f(x^{(i)}) - y^{(i)}]^2 \tag{11.14a}$$

$$J(b, w) = \frac{1}{2m} \sum_{i=0}^{m} [(b + wx^{(i)}) - y^{(i)}]^2 \tag{11.14b}$$

The partial derivatives of J become

$$\frac{\partial J(b, w)}{\partial b} = \frac{1}{m} \sum_{i=0}^{m} [f(x^{(i)}) - y^{(i)}]$$

$$\frac{\partial J(b, w)}{\partial w} = \frac{1}{m} \sum_{i=0}^{m} [f(x^{(i)}) - y^{(i)}] x^{(i)}$$

And equations 11.11 become

$$b_{new} = b - \alpha \frac{1}{m} \sum_{i=0}^{m} [f(x^{(i)}) - y^{(i)}] \tag{11.15a}$$

$$w_{new} = w - \alpha \frac{1}{m} \sum_{i=0}^{m} [f(x^{(i)}) - y^{(i)}] x^{(i)} \tag{11.15b}$$

Example 11.1 Python program for linear regression
For this example a deviation from a straight line equation is created with the random number generator. Consider the equation for the straight line given by

$$y = 2 + 4x \tag{11.16}$$

To generate a set of data points, a straight line is contaminated with random noise using the random number generator from the library **random**, the set of points (x, y) is generated as (import the libraries **numpy** and **random**)

```
from random import random, randint
import numpy as np
y = np.zeros(100)
```

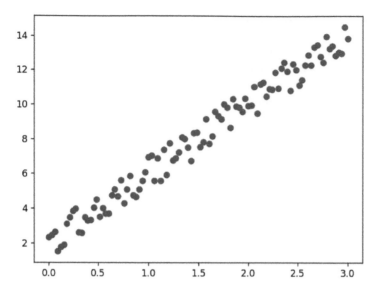

FIGURE 11.6: Data points.

```
x = np.linspace(0, 3, 100)
for i in range(len(x)):
    y[i] = 2 + 4*x[i] + (-1)**randint(0,1)*random()
```

A plot of these data points is shown in Figure 11.6. Now, the learning rate α, and the initial values for b and w are given as

```
alpha = 0.001
w = 0
b = 0
```

To implement equations 11.15 the value of m is evaluated as

```
m = len(y)
```

A while loop is used for the computation

```
while (True):
```

The first step inside the loop is to initialize the sums in equations 11.15 to 0:

```
sum0 = 0
sum1 = 0
```

A `for` loop is used to compute the sums

```
for j in range(len(y)):
    sum0 = sum0 + (b + w*x[j] - y[j])
    sum1 = sum1 + (b + w*x[j] - y[j])*x[j]
```

The new values for `b` and `w` are

```
bn = b -(2*alpha/m)*sum0
wn = w -(2*alpha/m)*sum1
```

The `while` loop is terminated when equations 11.13 are satisfied:

```
if abs(b-bn) < 0.00001 and abs(w-wn) < 0.00001:
    break
```

But if any of the inequalities do not hold, then the `b` and `w` values are updated:

```
else:
    b = bn
    w = wn
```

The complete `while` loop is

```
while (True):
    sum0 = 0
    sum1 = 0

    for j in range(len(y)):
        sum0 = sum0 + (b + w*x[j] - y[j])
        sum1 = sum1 + (b + w*x[j] - y[j])*x[j]

    bn = b -(2*alpha/m)*sum0
    wn = w -(2*alpha/m)*sum1

    if abs(b - bn) < 0.00001 and \
       abs(w - wn) < 0.00001:
        break

    else:
        b = bn
        w = wn
```

Finally, the total number of iterations and the values for `b` and `w` are displayed and the straight line corresponding to the computed values (solid line) and the original straight line (dotted line) are plotted together with the data points:

```
    print('The total number of iterations is ', i)
    print('Final')
    print('b = ', b)
    print('w = ', w)
    plt.plot(x, b + w*x, color = 'red')
    plt.plot(x, 2 + 4*x, linestyle = 'dotted')
```

The complete script is

```
"'' This is script Example_11_1.py
This program implements linear regression.'"

# Import the necessary libraries and functions
from random import random, randint
import numpy as np
import matplotlib.pyplot as plt

# The data points are generated:
y = np.zeros(100)
x = np.linspace(0, 3, 100)
for i in range(len(x)):
    y[i] = 2 + 4*x[i] + (-1)**randint(0,1)*random()

# Parameters are initialized:
alpha = 0.001
b = 0
w = 0
e = 1e-6 # Value of epsilon.

# Value of m is computed:

m = len(y)

# A while loop is used for the computation of b and w.
# Gradient descent algorithm is done in the while loop:

while (True):
    # Initialize the sums to 0:
    sum0 = 0
    sum1 = 0

# Script continues at next page.
```

```
# Script continued from previous page.

  # A for loop is used to compute the sums
  for j in range(len(y)):
     sum0 = sum0 + (b + w*x[j] - y[j])
     sum1 = sum1 + (b + w*x[j] - y[j])*x[j]

  # The new values for b and w are:

  bn = b -(2*alpha/m)*sum0
  wn = w -(2*alpha/m)*sum1

  # The conditions to terminate the loop are:
  if abs(b - bn) < e and abs(w - wn) < e:
     break

  # If conditions do not hold, update b and w as:

  else:
     b = bn
     w = wn
# Print No. of iterations, and final results.
# Plot the data and results:
print('The total number of iterations is ', i)
print('Final')
print('b = ', b)
print('w = ', w)
plt.scatter(x, y)
plt.plot(x, b + w*x, color = 'red')
plt.plot(x, 2 + 4*x, linestyle = 'dotted')
plt.show()
```

A run of the script produces:

```
The total number of iterations is 4398
Final
b = 2.103592476950139
w = 3.976179042022216
```

A plot for the data points and the two straight lines is shown in Figure 11.7. The dotted line is the original straight line. Due to the introduced random noise the coefficients cannot be obtained exactly.

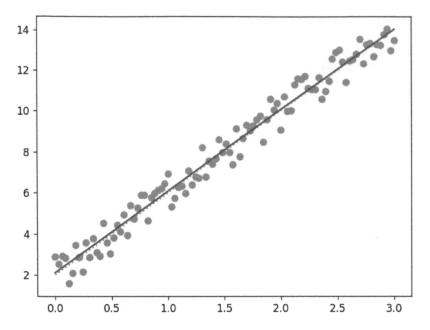

FIGURE 11.7: Data points and linear regression using the gradient descent.

11.3   Multivariate Regression

For the multivariate regression problem equations 11.15, repeat for each and every one of the features as shown here:

$$b_{new} = b - \alpha \frac{1}{m} \sum_{i=0}^{m} (y^{(i)} - 1) \qquad (11.17a)$$

$$w_{k,new} = w_k - \alpha \frac{1}{m} \sum_{i=0}^{m} (y^{(i)} - x^{(i)})x^{(i)} \qquad (11.17b)$$

$$\vdots$$

for $k = 1, \ldots, n$

11.3.1   Feature Scaling for the Multivariate Case

There are two techniques for feature scaling for the case of multivariate linear regression. These techniques are: feature scaling and mean normalization.

Feature scaling consists in dividing the input values by the range, that is, the maximum value minus the minimum value of the input variable. The result is a variable in a new range between 0 and 1.

Mean normalization consists in subtracting the average value for a feature from the values for that variable, resulting in a new average value for the feature that is equal to zero.

Both techniques are implemented in the equation:

$$x_i = \frac{x_1 - \mu_i}{s_1} \tag{11.18}$$

where μ_i is the average for the values for feature i and s_i is either the range of values (max - min) or s_i is the standard deviation. Note that both cases for s_i give different results.

11.4   The Normal Equation

So far the approach to solve linear regression problems has been through the use of optimization. There is another approach to solve some linear regression problems. The technique is known as the *Normal equation* and is an analytical solution. To introduce the concept consider the data in Table 11.2 and suppose for simplicity that there are only four training examples, the ones shown in the table. The features shown there can be arranged in a matrix that is called the *design matrix* as

$$X = \begin{bmatrix} 1 & 2000 & 3 & 4 & 12 \\ 1 & 1527 & 4 & 4 & 15 \\ 1 & 1858 & 3 & 4 & 14 \\ 1 & 987 & 2 & 2 & 17 \end{bmatrix} \tag{11.19}$$

The y values can be arranged in a column vector as

$$y = \begin{bmatrix} 200 \\ 150 \\ 250 \\ 89 \end{bmatrix} \tag{11.20}$$

Note that the design matrix is not a square matrix. The analytic solution given by the normal equation is

$$W = (X^T X)^{-1} X^T y \tag{11.21}$$

There are some pros and cons for this normal equation. Some of them are listed in Table 11.4 and they do not need further explanation.

TABLE 11.4:   Comparison between optimization and the normal equation

Optimization	Normal equation
It needs many iterations	No need to iterate.
It works well when n is large	It is very slow for a large n.
$O(n^2)$	$O(n^3)$ because of the inverse calculation.

An issue that arises sometimes is the fact that $X^T X$ may be non invertible. This problem can be alleviated by looking for features that are either linearly dependent or very close to be linearly dependent. In this case those linearly dependencies are eliminated by deleting the features that are linearly dependent. In some other cases it is necessary to delete some features because there are too many of them thus reducing the size of the design matrix.

Example 11.2 Linear regression using the normal equation
The normal equation to obtain the linear regression is performed. As in the previous example the data points are generated by adding noise to a straight line. The first step is to import the necessary libraries. This is done with

```
import matplotlib.pyplot as plt
import numpy as np
from random import random, randint
```

Next, the data point are generated with random deviation for the straight line

```
y = np.zeros(100)
x = np.linspace(0, 3, 100)
for i in range(len(x)):
    y[i] = 2 + 4*x[i] + (-1)**randint(0,1)*random()
```

The matrix X, according to equation 11.19 has only two columns and the first column is composed of 1's and the second column has the features $x^{(i)}$. The matrix X can be obtained with a concatenation using

```
X = np.c_[np.ones((-1, 1)), x]
```

where the method c_[ ] concatenates the arrays inside the brackets. The resulting array is a numpy array. For numpy arrays a and b, the dot product can be implemented as a.dot(b). The inverse matrix for an array A can be implemented with np.linalg.inv(A). Then, equation 11.19 can be implemented with

```
W = np.linalg.inv(X.T.dot(X)).dot(X.T).dot(y)
```

In Python code the equation is

```
W = np.linalg.inv(X.T.dot(X)).dot(X.T).dot(y)
```

Finally, the result for W[0] and W[1] are obtained with print('W', W) and the data points and the straight line obtained with the W values can be plotted with

```
print('W', W)
plt.scatter(x, y)
plt.plot(x, W[0] + x*W[1], color = 'blue')
plt.plot(x, 2 + 4*x, color = 'black')
plt.show()
```

The complete script is

```
"' This is script Example_11_2.py
This program uses the normal equation for linear regression."'

# Import the necessary libraries
import matplotlib.pyplot as plt
import numpy as np
from random import random, randint

# Data points generated with random deviation
# for the straight line:
y = np.zeros(100)
x = np.linspace(0, 3, 100)
for i in range(len(x)):
    y[i] = 2 + 4*x[i] + (-1)**randint(0,1)*random()

# Matrix X can be obtained with a concatenation using:
X=np.c_[np.ones((100, 1)),x]

# The matrix W is then:
W = np.linalg.inv(X.T.dot(X)).dot(X.T).dot(y)

# W[0] and W[1] are obtained with:
print('W', W)

# The data points and the straight line obtained
# with W are plotted

# Script continues at next page.
```

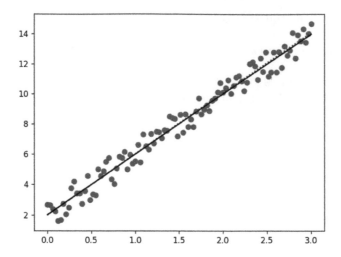

FIGURE 11.8: Data points and linear regression using the normal equation.

```
# Script continued from previous page.

plt.scatter(x, y)
plt.plot(x, W[0] + x*W[1], color = 'blue')
plt.plot(x, 2 + 4*x, color = 'black')
plt.show()
```

A run produces the values for W as

    W [1.98119633 4.04120188]

The data points generated are shown in Figure 11.8 together with the straight line whose parameters are: W [1.98119633 4.04120188] and the plots are shown in Figure 11.8.

11.5　The Package `scikit-learn`

The package `scikit-learn` has been designed for machine learning applications and it supports supervised and unsupervised learning. It provides tools for model fitting, data preprocessing, model selection, model evaluation, and many other utilities. It can be installed with the `pip` tool as shown in Appendix A. Several examples in the chapter show how to use it. The `scikit-learn` site provides details on the use of it. The site is `scikit-learn.org`. In this chapter the linear regression and the logic regression methods are used. As the first example of the use of scikit-learn, the linear regression of Example 11.1 is implemented. The library `scikit-learn` is imported as `sklearn`.

Example 11.3 Linear regression using `scikit-learn`

The data generated in Example 11.1 is used with `scikit-learn`. In the example a linear regression was implemented in Python from scratch. The first thing to do is to import the necessary libraries to plot using `matplotlib.pyplot`, to generate the vectors using the `numpy` package, and to use `scikit-learn`. The keyword to import `scikit-learn` is `sklearn`. The method used for linear regression in `scikit-learn` is `LinearRegression` which is in the sublibrary `linear_model`. Then,

```
from sklearn.linear_model import LinearRegression
import matplotlib.pyplot as plt
import numpy as np
from random import random, randint
```

The data points are generated in the same way and stored in vectors X and y. This is done with

```
y = np.zeros(100)
x = np.linspace(0, 3, 100)
for i in range(len(x)):
    y[i] = 2 + 4*x[i] + (-1)**randint(0,1)*random()
```

The arrays X and y have to be reshaped as column arrays to be accepted by `sklearn`,

```
x = np.reshape(x, (-1,1))
y = np.reshape(y, (-1,1))
```

The model is defined as

```
lin_reg = LinearRegression()
```

The actual linear regression is implemented with the `fit` method as

```
lin_reg.fit(x, y)
```

The `lin_reg` method now has produced the y-axis intercept and the slope of the straight line. These parameters are obtained with

```
b = lin_reg.intercept_
```

```
w = lin_reg.coef_
print('The y-axis intercept and the slope are:', b, w)
```

Finally, the data points and the straight line from the linear regression are plotted and displayed with

```
plt.scatter(x, y)
plt.plot(x, b + w*x, color = 'black')
plt.show()
```

In all the examples above, the training of the linear regression has been done with all the samples in the data. A better approach is to separate some of the data for the training and use the rest of the data for testing the linear regression algorithm. This task can be accomplished with the method `train_test_split` available in the library `sklearn.model_selection`. The training part can be a fraction of the data and the rest of it can be used for testing. The format is

```
train_test_split(x, y, train_size, test_size, random_state = 23)
```

For example, a good choice for `train_size` may be 90% of the data for training and 10% for testing. The value for the `test_size` is optional and can be omitted. The choice `random_state = 23` is for a random selection of the data at each run. The splitting of the data is

```
x_train, x_test, y_train, y_test = \
    train_test_split(x, y, train_size = 0.8, random_state=23)
```

The fitting is done with

```
lin_reg.fit(x_train, y_train)
```

The squared error can be obtained with the method `r2_score` from the library `sklearn.metrics`. The library is imported

```
from sklearn.metrics import r2_score
```

The script for this part is

```
from sklearn.model_selection import train_test_split
from sklearn.metrics import r2_score

# Split the data for training and testing:
x_train, x_test, y_train, y_test = train_test_split \
    (x, y, train_size = 0.8, random_state = 23)
```

```
lin_reg = LinearRegression()
lin_reg.fit(x_train, y_train)

# Coefficients b and w
bs = lin_reg.intercept_
ws = lin_reg.coef_

# The squared error is computed. The predicted values are:
yp = ws*x_train + bs
# Squared error
sqs = r2_score(y_train, yp)

# The squared error is printed:
print('The value of r^2 is: ' , sqs)

# The squared error without training splitting is:
y = wx + b
sqs = r2_score(y, yp)

# The squared error is printed:
print('The value without splitting of r^2 is: ' , sqs)
```

The complete script is:

```
# Example_11_3.py Linear regression with scikit-learn.

# The required libraries are imported:
from sklearn.linear_model import LinearRegression
from random import randint, random
import matplotlib.pyplot as plt
import numpy as np

# The data points are generated:
y = np.zeros(100)
x = np.linspace(0, 3, 100)
for i in range(len(x)):
    y[i] = 2 + 4*x[i] + (-1)**randint(0,1)*random()

# Script continues at next page.
```

```
# Script continued from previous page.

# The arrays x and y have to be reshaped as column arrays.
# This is required by sklearn:

x = np.reshape(x, (-1,1))
y = np.reshape(y, (-1,1))

# The model is

lin_reg = LinearRegression()

# The linear regression is implemented with the fit method as:

lin_reg.fit(x, y)

# The y-axis intercept and the slope are:

b = lin_reg.intercept_
w = lin_reg.coef_

# Linear regression using part of data for training
# and part for testing.
# The necessary libraries are imported:
from sklearn.model_selection import train_test_split
from sklearn.metrics import r2_score

# The data is splitted, 80% for trainig and 20% for testing:
x_train, x_test, y_train, y_test = \
    train_test_split(x, y, train_size = 0.8, random_state=23)

# The model is created:
lin_reg = LinearRegression()
lin_reg.fit(x_train, y_train)
# Coefficients b and w and m
bs = lin_reg.intercept_
ws = lin_reg.coef_

# Script continues at next page.
```

```
# Script continued from previous page.

# The squared error is computed. The predicted values are:
yps = ws*x_train + bs
# Squared error
sqs = r2_score(y_train, yps)

# Displaying the results
# The coefficients b, w, and the squared error
# when the data are splitted are printed:
print('The y-axis intercept and the slope are: \
      \n', b[0], w[0][0])
print('The value of r∧2 is: ' , sqs)

# The squared error without training splitting is:
yp = w*x + b
sqs = r2_score(y, yp)

# The squared error is printed:
print('The value without splitting of r∧2 is: ' , sqs)

# The coefficients bs, ws, and the squared error are printed.
# The coefficients b0, w1, and the squared error are printed:
print('The y-axis intercept and the slope when data \
      is splitted are:\n', bs[0], ws[0][0])
print('The value of r∧2 is: ' , sqs)

# The data points and the straight line from
# the linear regression are plotted:
plt.scatter(x, y)
plt.plot(x, b + w*x, color = 'black')
plt.plot(x, bs + ws*x, color = 'red')
plt.show()
```

A run for this script produces the values for b, w_1, $b0s$ and w_1s for both cases are given by

```
The y-axis intercept and the slope are,
respectively:

1.8485305132324061 4.066335371892183
The value of r∧2 is: 0.9805030083081524
```

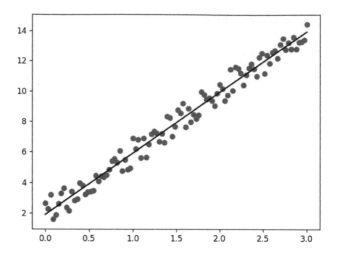

FIGURE 11.9: Data points and linear regression straight line from scikit-learn.

The y-axis intercept and the slope, when data is splitted, are, respectively:

```
1.82468152 4.00090227
The value of r^2 is: 0.9787833851121427
```

The data points generated are shown in Figure 11.9 together with the straight a line whose parameters are:

```
b = 1.8485305132324061
w = 4.066335371892183
```

It can be seen that the linear regression obtained with `scikit-learn` is very close to the results obtained in examples, 11.1 and 11.2. In this example the model definition, the `LinearRegression` method and the `fit` method were used.

11.6   Polynomial Regression

Linear regression has been used so far, but this is not necessarily the best option. A better choice for the hypothesis function may be a quadratic polynomial and in some cases a cubic one, such as

$$f_\theta(x) = b + w_1 x + w_2 x^2 \qquad (11.22)$$

$$f_w(x) = b + w_1 x + w_2 x^2 + w_3 x^3$$

It may be the case that the exponent is a fractionary one such as in the equation

$$f_w(x) = b + w_1 x + w_2 \sqrt{x}$$

Polynomial regression may be a better choice for the hypothesis function but it may be computationally more expensive.

Example 11.4 Polynomial regression using `scikit-learn`.
Consider the data generated by adding noise to a third-degree polynomial given by

$$f(x) = 2 - 3x + 2x^3$$

To generate the data for this case, first import the necessary libraries and methods and then generate a vector for x values as

```
# Import the necessary libraries:
import matplotlib.pyplot as plt
from random import random, randint
import numpy as np
from sklearn.preprocessing import PolynomialFeatures
from sklearn.linear_model import LinearRegression
```

The values for x are from $x = -3$ to $x = 3$. The y-values are initialized as 0:

```
x = np.linspace(-3, 3, 100)
y = np.zeros(100)
```

The data is generated by adding noise to this polynomial as

```
for i in range(len(x)):
    y[i] = 2*x[i]**3 - 3*x[i] + 2 + 10*(-1)**randint(0,1)*random()
```

`scikit-learn` requires column vectors. They are reshaped with

```
x = np.reshape(x, (-1, 1))
y = np.reshape(y, (-1, 1))
```

The model for polynomial regression is created with the method `PolynomialFeatures` and the fitting is done with

```
poly_features = PolynomialFeatures(degree = 3, \
```

```
        include_bias = False)
Xpoly = poly_features.fit_transform(x)
```

Now fit a linear regression model

```
    lin_reg = LinearRegression()
    lin_reg.fit(Xpoly, y)
```

The y-axis intercept is obtained with `lin_reg.intercept_` and the coefficients of the cubic polynomial are obtained with `lin_reg.coef_`. The coefficients of the polynomial regression are then

```
    b = lin_reg.intercept_[0]
    w1 = lin_reg.coef_[0][0]
    w2 = lin_reg.coef_[0][1]
    w3 = lin_reg.coef_[0][2]
```

The resulting polynomial can be obtained with these coefficients as

```
    yp = b + w1*x + w2*x**2 + w3*x**3
```

And the squared error is obtained with

```
    sqs = r2_score(yp, y)
    print('sqs = ', sqs)
```

Finally, the plot of the y-points and the yp-points is done with

```
    plt.scatter(x,y)
    yp = b+w1*x+w2*x**2+w3*x**3
    plt.plot(x, yp)
```

The complete script is

```
# Example_11_4.py Polynomial regression with scikit-learn.
# This example implements polynomial regression.

# Import the necessary libraries:
import matplotlib.pyplot as plt
from random import random, randint
import numpy as np
from sklearn.preprocessing import PolynomialFeatures
from sklearn.linear_model import LinearRegression

# Script continues at next page.
```

```
# Script continued from previous page.

# The values for x are from x = -3 to x = 3.
# The y-values are initialized as 0:

x = np.linspace(-3, 3, 100)
y = np.zeros(100)

# Noise is added to the cubic polynomial:
for i in range(len(x)):
    y[i]=2*x[i]**3 - 3*x[i] +2+10*(-1)**randint(0,1)*random()

# Reshape the vectors as column vectors:
x = np.reshape(x, (-1, 1))
y = np.reshape(y, (-1, 1))

# The model for polynomial regression and the fitting is:
poly= PolynomialFeatures(degree = 3, include_bias = False)
Xpoly = poly.fit_transform(x)

# Now fit a linear regression model
lin_reg = LinearRegression()
lin_reg.fit(Xpoly, y)

# The y-axis intercept is obtained with lin_reg.intercept_[0]
# The coefficients of the cubic polynomial
# are obtained with lin_reg.coef_.
# The coefficients of the polynomial regression are:

b = lin_reg.intercept_[0]
w1 = lin_reg.coef_[0][0]
w2 = lin_reg.coef_[0][1]
w3 = lin_reg.coef_[0][2]

# The results are printed out:
print('b = ', b)
print('w1 = ', w1)
print('w2 = ', w2)
print('w3 = ', w3)
yp = b + w1*x + w2*x**2 + w3*x**3

# Script continues next page.
```

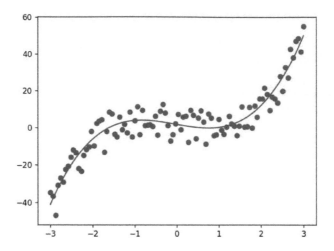

FIGURE 11.10: Polynomial regression.

```
# Script continued from previous page.

# The squared error is obtained with
from sklearn.metrics import r2_score
sqs = r2_score(yp, y)
print('sqs = ', sqs)

# The plot of the y-points and the yp-points is done
plt.scatter(x,y)
plt.plot(x, yp)
plt.show()
```

A run produces a squared error and the coefficients

```
sqs = 0.8566338231859216
b = 1.3815450427785627
w1 = -3.3138517549947593
w2 = -0.07957952825220471
w3 = 2.010727385384017
```

A comparison with the original polynomial shows that the independent coefficient deviates from the value 2 to 1.3815450427785627, the coefficient of x deviates from -3 to -3.3138517549947593, the coefficient of x∧2 goes from 0 to -0.07957952825220471, and finally, the coefficient of x∧3 deviates from 2 to 2.010727385384017. The original values cannot be obtained due to the added random noise. Furthermore, each run will produce different results due to the random noise that was added. A plot showing the data points and the cubic polynomial is shown in Figure 11.10.

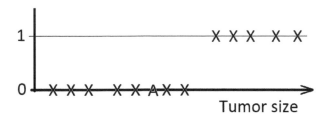

FIGURE 11.11: Data for breast tumor patients.

11.7   Classification with Logistic Regression

Classification is another technique for supervised machine learning. In this type of learning, the output takes on discrete values. Examples of a classification problem are:

- Is an email spam?

- Is a bank transaction fraudulent?

- Is a tumor malignant?

- Is a person diabetic?

These are examples whose answer can be either yes or no. Denoting the answer with the letter y, it is readily seen that y can only have two values: yes or no, true or false, 1 or 0. This type of supervised learning algorithm whose output can take on only two values is known as *binary classification*.

As an example, a plot of medical records is shown in Figure 11.11 where the crosses reveal a malignant tumor and the circles a benign one. The feature in the horizontal axis refers to the size of the tumors and the vertical axis has marks at 0 and 1. The value of 1 has been chosen for the malignant tumors and 0 for the benign ones. A malignant tumor is a tumor that is harmful and dangerous and a benign tumor is a tumor that may be harmless but it requires attention. For a person whose tumor has a size indicated by the letter A, it can be concluded that the tumor is benign. In this plot there are eight tumors that are benign and five tumors that are malignant. This is an example of a classification problem. It turns out that in classification problems, sometimes there are more than two possible values for the output. As a concrete example, maybe there is more than one type of cancer. So, the task is to try to predict a discrete value output zero, one, two, or three, where zero may mean benign, one may mean type-one cancer, and two mean a second type of cancer, and three may mean a third type of cancer.

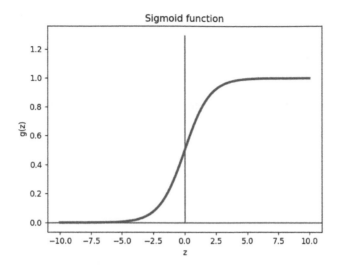

FIGURE 11.12: Sigmoid function.

11.7.1   Logistic regression

The classification problem cannot be solved using linear regression, instead a different hypothesis function must be used. For this purpose, the *sigmoid* function is introduced. This function is defined by

$$g(z) = \frac{1}{1 + e^{-z}} \tag{11.23}$$

The sigmoid function is plotted in Figure 11.12. It can be seen that the function values are between 0 and 1 and they are equal to 1 for $z \gg 0$ and to 0 when $z \ll 0$. Recalling linear regression, the variable z can be written as

$$z = b + W^T x \tag{11.24}$$

In the multivariate case W is a vector. The function can then be written as

$$f_W(x) = g(b + W^T x) \tag{11.25}$$

which is the same as

$$f_W(x) = \frac{1}{1 + e^{-(b + W^T x)}} \tag{11.26}$$

This function is known as a threshold classifier because if its value is much greater than zero it produces an output equal to 1, and if the value is much smaller than 0 it produces an output of zero. A plot of the function for a single feature x and values of $b = -3.6$ and $w = 0.09$ is shown in Figure 11.13.

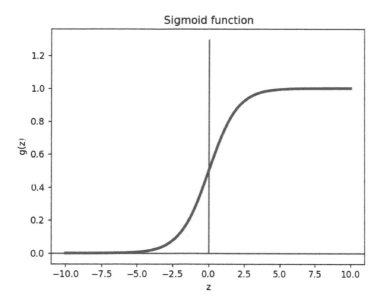

FIGURE 11.13: Sigmoid function with $b = -3.6$ and $w = 0.09$.

The cost function for logistic regression is defined as

$$J(W, b) = \frac{1}{m} \sum_{i=1}^{m} \frac{1}{2} [f_{W,b}(x^{(i)}) - y^{(i)}]^2 \qquad (11.27)$$

The loss function is defined by

$$L(f_{W,b}(x^{(i)}), y^{(i)}) = \frac{1}{2} [f_{W,b}(x^{(i)}) - y^{(i)}]^2 \qquad (11.28)$$

Using the sigmoid function, it can be shown that the loss function is given by

$$L(f_{W,b}(x^{(i)}), y^{(i)}) = -y^{(i)} \log(f_{W,b}(x^{(i)})) - (1 - y^{(i)}) \log(1 - f_{W,b}(x^{(i)})) \qquad (11.29)$$

and so the cost function is

$$J(W, b) = -\frac{1}{m} \sum_{i=0}^{m} L(f_{W,b}(x^{(i)}), y^{(i)}) \qquad (11.30)$$

Finally, the cost function is given by

$$J(W, b) = -\frac{1}{m} \sum_{i=0}^{m} [y^{(i)} \log(f_{W,b}(x^{(i)})) - (1 - y^{(i)}) \log(1 - f_{W,b}(x^{(i)}))] \qquad (11.31)$$

where

$$f_{W,b} = g(X, W, b) \tag{11.32}$$

The output y can only be 0 or 1. Thus, when $y = 1$, only the first term in equation 11.31 holds, and when the value of $y = 0$ the second term holds. The first term is valid when the output is equal to unity. For this first term, when h is close to zero, the value of $-\log(h)$ is very large and thus, that value does not minimize the cost function. The second term is valid when $y = 0$ and when it is close to unity the function $\log(0)$ approaches infinity which does not minimize the cost function.

To optimize this cost function, gradient descent can be used. The equations for gradient descent are repeated here for convenience:

$$|w_{j,new} - w| < \epsilon \tag{11.33a}$$

$$|b_{new} - b| < \epsilon \tag{11.33b}$$

The partial derivatives of the cost function J are:

$$\frac{\partial J(W, b)}{\partial w_j} = \frac{1}{m} \sum_{i=0}^{m} (f_{W,b}(x^{(i)}) - y^{(i)}) x^{(i)} \tag{11.34a}$$

$$\frac{\partial J(b, w)}{\partial b} = \frac{1}{m} \sum_{i=0}^{m} (f_{W,b}(x^{(i)}) - y^i) \tag{11.34b}$$

The equations to iterate the w and b parameters are

$$w_{j,new} = w_j - \alpha \frac{\partial}{\partial w_j} \sum_{i=0}^{m} (f_{W,b}(x^{(i)}) - y^{(i)}) x^{(i)} \tag{11.35a}$$

$$b_{new} = b - \alpha \frac{1}{m} \sum_{i=0}^{m} (f_{W,b}(x^{(i)}) - y^{(i)}) \tag{11.35b}$$

These equations are identical to equations 11.15. The only difference is in the function $f(x)$ which in logistic regression is the sigmoid function.

Example 11.5 Logistic regression for a diabetes analysis

There is a set of parameters that are considered risky when evaluating a potential diabetic patient. In this example, eight risk factors are considered to assess if a patient is diabetic. In this example only a parameter is used to assess the patient, then it is required to analyze the data for a set of patients with a certain level of blood pressure, some of which are diagnosed as positive. Develop a learning algorithm that indicates if a patient may be positive to diabetes. The data is available at the site

TABLE 11.5: Diabetes data

x0	x1	x2	x3	x4	x5	x6	x7	Outcome
6	148	72	35	0	33.6	0.627	50	1
1	85	66	29	0	26.6	0.351	31	0
8	183	64	0	0	23.3	0.672	32	1
1	89	66	23	94	28.1	0.167	21	0
0	137	40	35	168	43.1	2.288	33	1
5	116	74	0	0	25.6	0.201	30	0

https://www.kaggle.com/datasets/kandij/diabetes-dataset

The file is `diabetes2.csv` and it is available in the data for `Diabetics prediction using logistic regression`. The data is shown in Table 11.5 for a few patients. In this example, a single feature is used and the one chosen is the blood pressure. As can be seen from the table this is not the only factor to declare a person as a diabetic one, but this example is with a single feature and the blood pressure level is the selected feature.

In Table 11.5, the features x0 to x7 are:

- x0   Pregnancies.

- x1   Glucose.

- x2   BloodPressure.

- x3   SkinThickness.

- x4   Insulin.

- x5   BMI (Body Mass Index).

- x6   DiabetesPedigreeFunction.

- x7   Age.

and the last column is the output or target with the name `Outcome`.

This example is done with Pandas to handle the arrays. The first step is to import the necessary libraries which are `pandas`, `matplotlib`, and `numpy`. This is done with

```
# Import the necessary libraries:
import pandas as pd
import numpy as np
import matplotlib.pyplot as plt
```

Next, the data file is read as a `csv` file. This action creates a Pandas data frame:

```
data = pd.read_csv('diabetes2.csv')
```

If the data is printed, it can be seen that the data has a heading, 768 rows and 9 columns. The example only uses the column corresponding to `BloodPressure`. The printing of the column corresponding to `BloodPressure` and the `Output` is displayed with

```
print(train[['BloodPressure', 'Outcome']].head())
```

and only these two columns and the first five rows are displayed.

```
   BloodPressure Outcome
0             72       1
1             66       0
2             64       1
3             66       0
4             40       1
```

Now, equations 11.35 are implemented. A value for the learning rate α and initial values for b and w are required. They are initialized as

```
b = 0.0
w = 0.0
alpha_b = 0.1
alpha_w = 0.001
```

The sigmoid function from equation 11.26 is defined in a method as

```
def sigmoid(x, w, b):
      return 1/(1 + np.exp(-(x*w + b)))
```

The partial derivatives from equation 11.34 are computed for w as

```
data['sigmoid'] = sigmoid(data['BloodPressure'], w_0, b_0)))
data['loss'] = (data['sigmoid'] \
   - data['Outcome'])*data['BloodPressure']
derivative_w = data['loss'.mean()
```

and for b as

```
data['sigmoid']=1/(1+np.exp(-(data['BloodPressure']*w+b)))
data['partial_loss'] = (data['sigmoid']-data['Outcome'])
derivative_w = data['loss'.mean()
```

$$w_{new} = w - \frac{\alpha}{m} X^T (g(X, W, b) - \vec{y}) \qquad (11.36)$$

The initial values for w, b, `alpha_w`, and `alpha_b` are

```
w_0 = 1
b_0 = -1
alpha_w = 0.001
alpha_b = 0.1
```

A new column 'sigmoid' is added to the data array as

$$data['sigmoid'] = sigmoid(data['BloodPressure'], w, b)$$

The gradient descent algorithm starts. A while loop is used for the iterations. It starts by initializing the index to 0:

```
i = 0
```

The while loop keeps working unless both equations 11.33 are satisfied; thus, the while loop instruction is:

```
while (True):
```

The first step is to compute the value of the sigmoid function for the initial values of w and b. This is done with (the following lines must have an indentation):

```
data['sigmoid'] = sigmoid(data['BloodPressure'], w_0, b_0)
```

A new column is added to the data array for the loss considering the w parameter. This is done with:

```
data['loss'] = (data['sigmoid'] -data['Outcome'])*data['BloodPressure']
```

Now the derivative of equation 11.34a can be obtained

```
derivative_w = data['loss'].mean()
```

The same procedure is followed for the derivative with respect to b. Then:

```
data['loss'] = (data['sigmoid']-data['Outcome'])
derivative_b = data['loss'].mean()
```

The new values of w and b are:

```
w_new = w_0 - alpha_w * derivative_w
b_new = b_0 - alpha_b * derivative_b
```

These new values are compared with the previous ones with equations 11.33 with a value for ϵ of 1×10^{-6} and using an if condition. If the condition is satisfied the loop ends, but otherwise update the values of w and b:

```
if np.abs(w_0 - w_new) < 1e-6 and np.abs(b_0 - b_new) < 1e-6:
    break
else:
    w_0 = w_new
    b_0 = b_new
```

This ends the algorithm. The values for w and b are printed

```
print('The value of w is: ', w_0)
print('The value of b is: ', b_0)
```

To check a result consider the value of blood pressure of 140 and see what the result produces the sigmoid:

```
x = 140
y = 1/(1 + np.exp(-(w_0*x + b_0)))
print('The value of the output is: ', y)
```

The result is 0.4750174379574634 which corresponds to a 0 and thus the person is not diabetic with a probability of 1 - 0.475 = 52.5%. The final values for w and b are:

```
The value of w is: 0.007434147354632805
The value of b is: -1.14079416168132
```

The complete script for this example is:

```
"' This is script Example_11_5.py
This program implements logistic regression."'

# Import the necessary libraries:
import pandas as pd
import numpy as np
import matplotlib.pyplot as plt

# The data file is read in as a csv file and
# it creates a Pandas data frame:
data = pd.read_csv('diabetes2.csv')

# The printing of the column corresponding to BloodPressure
# and the Output:
print(data[['BloodPressure', 'Outcome']].head())

# Equations 11.35 are implemented.
# The learning rates alpha_b and alpha_w and initial values
# for b and w are required. They are initialized as:

b = 0.0
w = 0.0
alpha_b = 0.1
alpha_w = 0.001

# The sigmoid function from equation 11.26 is defined
# in a method:
def sigmoid(x, w, b):
      return 1/(1 + np.exp(-(x*w + b)))

# The initial values for w, b, alpha_w, and alpha_b are

w_0 = 1
b_0 = -1
alpha_w = 0.001
alpha_b = 0.1

# A new column 'sigmoid' is added to the data array as
data['sigmoid'] = sigmoid(data['BloodPressure'],w, b)

# Script continues next page.
```

```
# Script continued from previous page.

# The gradient descent algorithm starts.
# A while loop is used for the iterations. It starts
# by initializing the index to 0:
i = 0

# The while loop keeps working unless both equations
# 11.33 are satisfied:

while(True):

    # The first step is to compute the value of the sigmoid
    # function for the initial values of w and b:
    data['sigmoid'] = sigmoid(data['BloodPressure'], w_0, b_0)

    # A new column is added to the data array for the loss
    # considering the w parameter:
    data['loss'] = (data['sigmoid']-data['Outcome']) \
        *data['BloodPressure']

    # The derivative of equation 11.34a is:
    derivative_w = data['loss'].mean()

    # The same procedure is followed for the derivative
    # with respect to b. Then:
    data['loss'] = (data['sigmoid']-data['Outcome'])
    derivative_b = data['loss'].mean()

    # The new values of w and b are:
    w_new = w_0 - alpha_w * derivative_w
    b_new = b_0 - alpha_b * derivative_b

    # Check equations 11.33 with a value for epsilon
    # of 1x10^(-6) and using an if condition.
    # If the condition is satisfied the loop ends,
    # but otherwise update the values of w and b:
    if np.abs(w_0-w_new)< 1e-6 and np.abs(b_0-b_new) < 1e-6:
        break
    else:
        w_0 = w_new
        b_0 = b_new

# Script continues next page.
```

```
# Script continued from previous page.

# This ends the algorithm. The values for w and b are printed
print('The value of w is: ', w_0)
print('The value of b is: ', b_0)

# To check a result consider the value of blood pressure
# of 140 and see the result:
x = 140
y = 1/(1 + np.exp(-(w_0*x + b_0)))
print('The value of the output is: ', y)
```

Example 11.6 Logistic regression with scikit-learn.
Example 11.5 is now solved with scikit-learn. The import of libraries and
the reading of the data array is done in the same way as in that example:

```
# Import the necessary libraries:
import pandas as pd
import numpy as np
import matplotlib.pyplot as plt
from sklearn.linear_model import LogisticRegression
```

Next, the data file is read in as a csv file. This action creates a Pandas data
frame:

```
data = pd.read_csv('diabetes2.csv')
```

From the previous example, it is known that the information for the target is
in the column Outcome. The model is created and the data is fitted:

```
model = LogisticRegression()
model.fit(X_data['BloodPressure'], X_data['Outcome'])
```

Finally, the b and the w parameters are printed out and compared with the
results in the previous example:

```
print(f'Parameter b: {model.intercept_}')
print(f'Parameter w: {model.coef_}')
```

When this script is run the result is:

```
Parameter b: [-1.14008318]
Parameter w: [[0.00742463]]
```

which is the same result as in Example 11.5. The complete script is

```
"' This is script Example_11_6.py
This program implements Example 11.5 using scikit-learn.'"

# Import the necessary libraries:
import pandas as pd
import numpy as np
import matplotlib.pyplot as plt
from sklearn.linear_model import LogisticRegression

# The data file is read. A Pandas dataframe is generated:
data = pd.read_csv('diabetes2.csv')

# The data frame is reshaped with:
# The target is in the column Outcome.
x_data = np.array(data['BloodPressure']).reshape(-1, 1)
y_data = np.array(data['Outcome'])

# The model is created and the data is fitted:
model = LogisticRegression()
model.fit(x_data, y_data)

# The parameters b and the w are printed out:
print(f'Parameter b: {model.intercept_}')
print(f'Parameter w: {model.coef_}')
```

Notice the simplification in the code using `scikit-learn`.

Example 11.7 Logistic regression with multiple features
In the previous examples, a single feature has been used even though there are
more features in the data provided. Now, more features are used. In addition,
the data is split into a training part and a testing part. The change appears
in the definition of the data to be fitted and in the coefficients produced. The
script starts importing the necessary libraries and methods:

```
import pandas as pd
import numpy as np
import matplotlib.pyplot as plt
from sklearn.linear_model import LogisticRegression
from sklearn.model_selection import train_test_split
```

The data is read in and the heading is printed:

```
data = pd.read_csv('diabetes2.csv')
print(data.head(10))
```

The data is split. The data has 9 columns, the last one is the `Outcome`. The testing size slice is set at 15% and the training slice is 85%.

```
X_train, X_test, y_train, y_test= train_test_split \
      (data.iloc[:, 0:8],data['Outcome'], test_size = 0.15)
```

The model is defined and the fitting is done:

```
model = LogisticRegression()
model.fit(X_train, y_train)
```

Now, the prediction is computed and printed. The prediction produces outputs according to the model calculated from the fitting:

```
prediction = model.predict(X_test)
print(prediction)
```

The score indicates what is the percentage of good results obtained with the model:

```
a = model.score(X_test, y_test)
print(a)
```

Finally, the parameters for the model are printed:

```
print(f'Parameter b: {model.intercept_}')
print(f'Parameter w: {model.coef_}')
```

The final script is

```
"' This is script Example_11_7.py
This program implements logistic regression with multiple
features.'"

# The necessary libraries and methods are imported:
import pandas as pd
import numpy as np
import matplotlib.pyplot as plt

# Script continues next page.
```

```
# Script continued from previous page.

from sklearn.linear_model import LogisticRegression
from sklearn.model_selection import train_test_split

# The data is read in and the heading is printed:
data = pd.read_csv('diabetes2.csv')
print(data.head(10))

# The data is split. The data has 9 columns,
# the last one is the Outcome.
#The training slice is 85%.
X_train, X_test, y_train, y_test= train_test_split \
     (data.iloc[:, 0:7],data['Outcome'], test_size = 0.15)

# The model is defined and the fitting is done:
model = LogisticRegression()
model.fit(X_train, y_train)

# The prediction is computed and printed:
prediction = model.predict(X_test)
print(prediction)

# The score is computed:
a = model.score(X_test, y_test)
print(a)

# Finally, the parameters for the model are printed:
print(f'Parameter b: {model.intercept_}')
print(f'Parameter w: {model.coef_}')
```

A run for this program produces (only the score and the coefficients are displayed here):

```
score = 0.7844827586206896
Parameter b: [-8.65761075]
Parameter w: [[ 0.1054894 0.03793538 -0.01325795 -0.00442582 \
      -0.00102874 0.0954708 0.92562988 0.01143644]]
```

In this run, the score indicates that 78.4% of the results are good as compared with the Outcome column. The coefficients indicate an exponent for the sigmoid function (the coefficients are displayed to three decimal digits). The exponent of equation 11.26 is given by:

$$z = -8.658 + W^T X$$

where

```
X = [x0, x1, x2, x3, x4, x5, x6 , x7]
```

and

```
W = [0.106, 0.038, -0.013, -0.004, -0.001, 0.095, 0.926, 0.0114]
```

Another run will produce a different result due to the random nature of the split method.

Example 11.8 Logistic regression with more features.

This example uses the data frame from the iris database and is a typical set of data for classification. The iris dataset is available in several data websites and libraries. The iris flower in this dataset involves four parameters: the petal length, the petal width, the sepal length, and the sepal width. In this example, the dataset includes 150 iris flowers of three different species: setosa, versicolor, and virginica. The dataset that is used comes from the **seaborn** library. The purpose of this example is to build a classifier to detect the type of iris based on the four characteristics of each type of iris flower. The example is solved with **scikit-learn**. The example shows some functions from **scikit-learn** that can be used to assess the learning algorithm. The first step is to import the libraries used such as **pandas**, **matplotlib**, **seaborn**, **scikit-learn** and **numpy**:

```
import pandas as pd
from sklearn.model_selection import train_test_split
import seaborn as sns
import matplotlib.pyplot as plt
```

Next, the dataset is loaded from **seaborn** and prints the headings:

```
iris = sns.load_dataset('iris')
print(iris.columns)
```

The printing is:

```
Index(['sepal_length', 'sepal_width', 'petal_length',
        'petal_width', 'species'])
```

The last column corresponds to the target column and it can be any of the three iris species, namely, **setosa**, **versicolor**, and **virginica**.

The methods in `scikit-learn` require that the output is in a numeric format, the names `setosa`, `versicolor`, and `virginica` change to 1, 2, 3, respectively. This is done with:

```
iris['species'] = iris['species'].replace('setosa':1, \
        'versicolor':2, 'virginica':3)
```

The data is split in a training part and a testing part with:

```
X_train, X_test, y_train, y_test = train_test_split \
        (iris[['sepal_length', 'sepal_width', 'petal_length', \
        'petal_width']], iris['species'],test_size = 0.3)
```

The model is defined. From `scikit-learn` the model for logistic regression must be imported and the fitting is implemented with:

```
from sklearn import linear_model
model = linear_model.LogisticRegression()
model.fit(X_train, y_train)
```

A prediction is run with `X_test` using the method `predict`. This method runs as:

```
prediction = model.predict(X_test)
print(prediction)
```

The score is another measure of the algorithm. A score of 1.00 indicates that is exact and a score less than 1.00 indicates the percentage of success. To obtain the score for the model use:

```
model.score(X_test, y_test)
print('score = ', model.score(X_test, y_test))
```

which in this example is `score = 0.977`; that is, the algorithm has an accuracy of 97.7%. The confusion matrix is a way to see graphically how many of the tests are correct and how many are wrong. This can be obtained with:

```
from sklearn.metrics import confusion_matrix
cm = confusion_matrix(y_test, prediction)
print(cm)
```

For this example, a confusion matrix is:

```
[[10  0  0]
 [ 0 14  0]
 [ 0  1 20]]
```

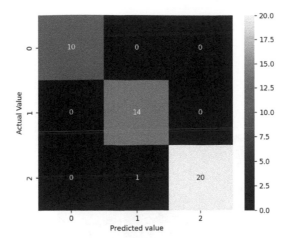

FIGURE 11.14: Confusion matrix plot.

which indicates that for the iris setosa, ten out of ten cases were predicted correctly. For the iris versicolor, fourteen out of fourteen were correctly predicted, and for the iris virginica, out of 22, 21 were correctly predicted and one was incorrect. The confusion matrix can be plotted with:

```
plt.figure(figsize = (5, 4))
sns.heatmap(cm, annot = True)
plt.xlabel('Predicted value')
plt.ylabel('Actual Value')
plt.show()
```

The plot produced is shown in Figure 11.14. This plot contains the same information displayed above in the confusion matrix. It is worth to point out that the confusion matrix changes for each run because the data from the run is produced by the logistic regression fit that is not a deterministic process. A second run produces the confusion matrix plot shown in Figure 11.15 and it is a different one as compared with the previous confusion matrix plot. Another plot of interest is a scatter plot of petal length vs. petal width that can be done with a **seaborn** scatter plot as

```
sns.scatterplot(x ='petal_length', y ='petal_width', \
     data = iris, hue = 'species')
```

The plot produced is shown in Figure 11.16. It is clearly seen that the setosa type is well separated from the virginica and the versicolor types. A decision boundary is well established for the setosa type but is not so well established

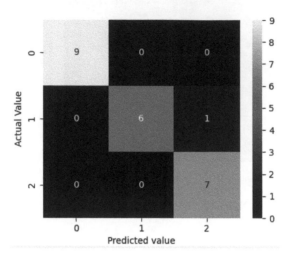

FIGURE 11.15: Confusion matrix plot for another run.

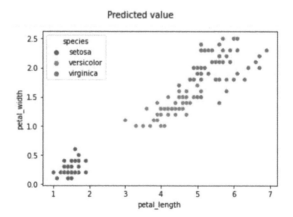

FIGURE 11.16: Petal length vs petal width for the iris dataset.

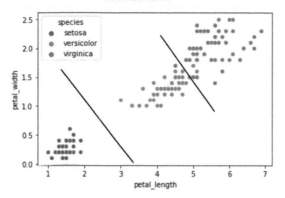

FIGURE 11.17: Decision boundaries added.

between the virginica and the versicolor types as can be seen in Figure 11.17 where the decision boundaries have been added outside of the script. The complete script is:

```
"' This is script Example_11_8.py
This program implements logistic regression with multiple
features.'"

# Import the necessary libraries:
import pandas as pd
import matplotlib.pyplot as plt
import seaborn as sns
from sklearn import linear_model
from sklearn.metrics import confusion_matrix
from sklearn.model_selection import train_test_split

# The data is read
iris = sns.load_dataset('iris')

# The target values are converted to numeric values:
iris['species'] = iris['species'].replace('setosa' == 1, \
     'versicolor' == 2, 'virginica' == 3)

# Script continues next page.
```

```
# Script continued from previous page.

# The data is split for training with 85%:
X_train, X_test, y_train, y_test = train_test_split \
      (iris[['sepal_length', 'sepal_width', 'petal_length', \
      'petal_width']], iris['species'],train_size = 0.85)

# The model is defined and the fitting is implemented with:
model = linear_model.LogisticRegression()
model.fit(X_train, y_train)

# A prediction is run for X_test:
# To test the model, the function predict is used as:
prediction = model.predict(X_test)
print(prediction)

# The model can be measured with the score method as:
a = model.score(X_test, y_test)
print(a)


# The confusion matrix is computed and printed:
cm = confusion_matrix(y_test, prediction)
print(cm)

# The confusion matrix is plotted with:
plt.figure(figsize = (5, 4))
sns.heatmap(cm, annot = True)
plt.xlabel('Predicted value')
plt.ylabel('Actual Value')

# The plot of petal width vs. petal lenght is produced:
sns.scatterplot(x ='petal_length', y ='petal_width', \
      data = iris, hue = 'species')

# The plots are displayed:
plt.show()
```

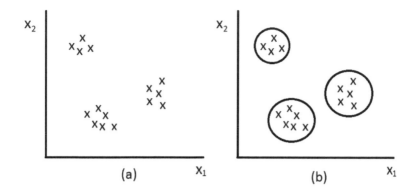

FIGURE 11.18: a) Unsupervised data, b) Clustered data.

11.8 Unsupervised Learning

Unsupervised learning does not require to know a right answer in advance or to know what type of data it is. For example, consider the data shown in Figure 11.18a for two features. It can be seen in the figure that no information is given for the data; that is, it is unlabeled. The only information that can be obtained from the data is the *clustering* shown in Figure 11.18b. Examples of unsupervised learning can be seen in market segmentation where companies group customers to sell or market different segments automatically. Clustering is only an example of the techniques available for unsupervised learning. *Visualization* algorithms is another technique for unsupervised learning where a great deal of complex and unlabeled data is fed to the algorithm and from there 2D and 3D representations can be extracted trying to preserve as much information as possible to preserve the structure of the data and possible detect unsuspected patterns. *Anomaly detection* is another important unsupervised technique used to detect anomalous instances in the data. Another common supervised task is called association rule learning and there the user looks for information associated with the data, for example, shoppers at a store on a given day are more inclined to shop a certain set of products, as opposed to what shoppers select on weekends. These are a few examples of the techniques available in unsupervised learning. In this chapter, only clustering will be covered.

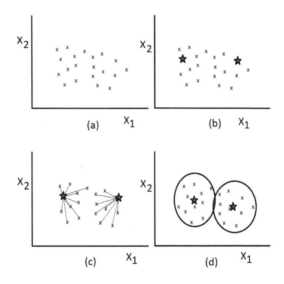

FIGURE 11.19: Clustered data.

11.9 Clustering Using k-means

The algorithm presented for clustering unlabeled data is known as **k-means**. It was proposed in 1957 by Stuart Lloyd from Bell Labs as a technique for pulse-code modulation. It was published outside of the company in 1957. Edward W. Forgy presented an equivalent algorithm in 1965; thus, the k-means algorithm is also known as the Lloyd-Forgy algorithm.

The k-means algorithm uses the concept of cluster. A *cluster* may be defined as a group of data points whose distances among themselves are small as compared to data points outside the cluster. To illustrate the mechanics of k-means, consider the data points shown in Figure 11.19a. It seems that there are two clusters. The clustering algorithm starts by assuming that one of the points in each cluster is the centroid for each of them as can be seen in Figure 11.19b. These two centroid points are indicated with a star. The next step is to take the distances from each of the proposed centroids to each and everyone of the points as shown in Figure 11.19c. The points closer to each centroid are assigned to the cluster of the centroid. The following step is to take the average of the points assigned to each centroid to give the location of the new centroid illustrated in Figure 11.19d. These four steps are summarized as:

1. Select one of the points of each cluster as a center of the cluster called the centroid of the cluster.

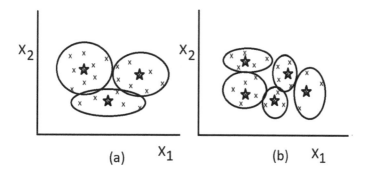

FIGURE 11.20: Clustered data.

2. Take the distances of the points to each of the proposed centroids.

3. Associate the points closer to each cluster point to the corresponding centroid.

4. Take the average of the position of each point associated with each centroid and designate this result as the new centroid position.

5. Repeat steps 2 and 3 until the centroid locations do not change.

6. The points associated with each centroid form each cluster.

Usually, the data points associated with each cluster are enclosed in some geometric shape such as a circle, square, or any other geometric shape, and they are called *blobs*. Sometimes it is not easy to "see" the clusters and thus, step No. 1 is a random choice. Consider Figure 11.20 where the data has been clustered into 3 and into 5 clusters. The question now is to know which is the best or optimum number of clusters. To answer this question a cost function has to be defined.

11.9.1 Cost Function

Assume that there are m examples or points and that there are K clusters. To define the first cost function, the above procedure to obtain the K centroids denoted as μ_k, $k = 0, \ldots, K\text{-}1$, is modified as follows:

1. Select one of the points of each cluster as a center of the cluster called the centroid μ_k of the cluster.

2. Take the distances of the points to each of the proposed centroids.

3. Associate the closer points to each cluster to the corresponding centroid.

4. Associate an index $c^{(k)}$ ($k = 0, \ldots, K$-1) to the point $x^{(k)}$.

5. Denote by $\mu_c^{(k)}$ the cluster index of the centroid associated with $x^{(k)}$.

6. Take the average of the position of each point associated with each centroid and assign this result as the new centroid position.

7. Repeat steps 2 to 6 until the centroid locations do not change.

8. The points associated with each centroid form each cluster.

9. Compute the Euclidean norm

$$||x^{(i)} - \mu_c^{(k)}||^2$$

10. The cost function $J(c^{(0)}, \ldots, c^{(K-1)}, \mu_0, \ldots, \mu_{K-1})$ is defined by

$$J(c^{(0)}, \ldots, c^{(K-1)}, \mu_0, \ldots, \mu_{K-1}) = \frac{1}{m} \sum_{i=0}^{m-1} ||x^{(i)} - \mu_c^{(i)}||^2$$

11. The minimization of this cost function is done over the parameters

$$c^{(0)}, \ldots, c^{(K-1)}, \mu_0, \ldots, \mu_{K-1}$$

It happens that the procedure of assignment of cluster centroids described above automatically minimizes the cost function. The cost function defined above is called the *distortion* function.

The package `scikit-learn` computes the inertia as a measure equivalent to the cost function and is defined by

$$\min_{\mu_j \in C} \sum_{i=0}^{m-1} ||x^{(i)} - \mu_j||^2$$

where C is the set of cluster points. Although both measures are different, they both produce optimum results.

The other factor that can decrease the cost function is the number of clusters. A plot of either the distortion or the inertia versus the number of clusters is shown in Figure 11.21 where it can be seen that there is not much improvement in the distortion (or the inertia) as the number of clusters is greater than 3. Sometimes this plot gives a good idea of the optimum number of clusters but some other times, the selection of the number of clusters is based upon the problem to solve, like in the case of T-shirt sizes, baby clothing, pants, and many more examples.

The package `scikit-learn` provides methods to implement k-means clustering. The following example generates the data and then implements the clustering using `scikit-learn`.

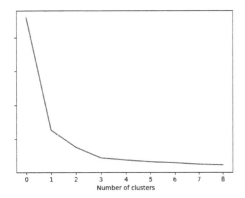

FIGURE 11.21: Plot of distortion vs. number of clusters.

Example 11.9 Cluster assignment.
Consider the data generated with the method `make_blobs` in the library
`sklearn.datasets` for 200 samples. Consider for the time being four clusters,
two features, the standard deviation for the clusters of 1.6, and the random
number generation of 50. The instructions to create the data set and plot the
points are:

```
from sklearn.datasets import make_blobs
# Create the data set:
dataset=make_blobs(n_samples=200, centers = 4,n_features=2, \
          cluster_std = 1.6, random_state = 50)
points = dataset[0]

# Plot the data set:
plt.scatter(dataset[0][:, 0], dataset[0][:, 1])
plt.xlabel('x1')
plt.xlabel('x2')
plt.show()
```

These instructions produce the data points shown in Figure 11.22. To find the
optimum value for the number of clusters, the values for the distortion and
inertia functions can be obtained as

```
from sklearn.cluster import KMeans
inertia = [ ]
distortion = [ ]
for i in range(1, 10):
     k_mean = KMeans(n_clusters = i, n_init = 'auto')
     k_mean.fit(points)
     inertia.append(k_mean.inertia_)
     distortions.append(sum(np.min(cdist(points,\
```

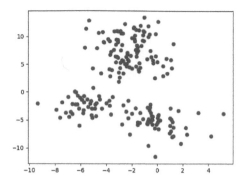

FIGURE 11.22: Data points produced by `make_blobs`.

```
    k_mean.cluster_centers_, 'euclidean' \
    ), axis = 1))/points.shape[0])

# Plot the distortion and inertia functions:
plt.plot([1,2,3,4,5,6,7,8,9], distortion)
plt.ylabel('Distortion')
plt.xlabel('Number of clusters')
plt.show()

plt.plot([1,2,3,4,5,6,7,8,9], inertia)
plt.ylabel('Inertia')
plt.xlabel('Number of clusters')
plt.show()
```

that produces Figures 11.23 and 11.24. It is readily seen that both measures provide the same results, namely, four clusters as the more appropriate num-

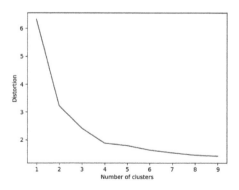

FIGURE 11.23: Plot of distortion vs. number of clusters.

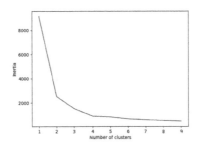

FIGURE 11.24: Plot of inertia vs. number of clusters.

ber. Now, it is necessary to create a k-means object with `n_clusters` equal to 4:

```
kmeans = KMeans(n_clusters = 4, n_init = 'auto')
```

and fit the data

```
kmeans.fit(dataset[0])
```

The cluster centroids are given by:

```
clusters = kmeans.cluster_centers_
print("The cluster points are: \n", clusters)
```

that are obtained as:

```
[[-2.40167949 10.17352695]
 [-5.56465793 -2.34988939]
 [ 0.05161133 -5.35489826]
 [-1.92101646 5.21673484]]
```

The data points and the centroids are plotted:

```
color0 = ['r', 'g', 'y', 'c', 'k']
y_kn = kmeans.fit_predict(points)
index = len(clusters)
for i in range(index):
    plt.scatter(points[y_kn == i,0], points[y_kn == i,1], \
    s = 50, color = color0[i])
for i in range(index):
    plt.scatter(clusters[i][0], clusters[i][1], \
    marker = '*', color='black', s=120)
plt.show()
```

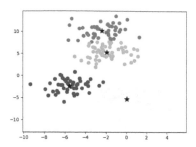

FIGURE 11.25: Data points with centroids.

The plot shown in Figure 11.25 is the result. The complete file is:

```python
# File Example_11_9.py
# Implements clustering using scikit-learn.

# Import the necessary libraries:
import matplotlib.pyplot as plt
from sklearn.cluster import KMeans
from sklearn.datasets import make_blobs
from scipy.spatial.distance import cdist
import numpy as np

# Create the data set:
dataset = make_blobs(n_samples=200, centers=4,n_features=2, \
cluster_std = 1.6, random_state = 50)
points = dataset[0]

# Plot the data set:
plt.scatter(dataset[0][:, 0], dataset[0][:, 1])
plt.xlabel('x1')
plt.ylabel('x2')
plt.show()

# Find the optimal number of clusters with the inertia
# and distortion functions:
inertia = [ ]
distortion = [ ]

# Script continues next page.
```

```
# Script continued from previous page.

for i in range(1, 10):
    k_means = KMeans(n_clusters = i, n_init = 'auto')
    k_means.fit(points)
    inertia.append(k_means.inertia_)
    distortion.append(sum(np.min(cdist(points,\
    k_means.cluster_centers_, 'euclidean'), \
    axis = 1))/points.shape[0])

# Plot the inertia values:
plt.plot([1,2,3,4,5,6,7,8,9], inertia)
plt.ylabel('Inertia')
plt.xlabel('Number of clusters')
plt.show()

# Plot the distortion values:
plt.plot([1,2,3,4,5,6,7,8,9], distortion)
plt.ylabel('Distortion')
plt.xlabel('Number of clusters')
plt.show()

# Run the KMeans method for n = 4:
kmeans = KMeans(n_clusters = 4, n_init = 'auto')
kmeans.fit(dataset[0])

# The cluster points are given by:
clusters = kmeans.cluster_centers_
print("The cluster points are:\n", clusters)

# The data points and the centroids are plotted with:
color0 = ['r', 'g', 'y', 'c', 'k']
y_kn = kmeans.fit_predict(points)
index = len(clusters)
for i in range(index):
    plt.scatter(points[y_kn == i,0], points[y_kn == i,1], \
    s = 50, color = color0[i])
for i in range(index):
    plt.scatter(clusters[i][0], clusters[i][1],
    marker = '*', color = 'black', s = 120)
plt.show()
```

TABLE 11.6: Python instructions in Chapter 11

Instruction	Description
coef_	It obtains the coefficients of a regression.
confusion_matrix	Graphical display of correct results.
cluster_centers	It obtains the cluster centers.
fit	It implements the fitting of an algorithm.
fit_predict	It fits and predicts the result of a fitting.
inertia_	Measure for a cost function.
inv	It obtains the inverse of a matrix.
intercept_	It obtains the y-axis intercept.
KMeans	Keras algorithm for clustering.
linalg	Library for linear algebra methods.
linear_model	It fits data to a linear model.
LinearRegression	sklearn method for linear regression.
LogisticRegression	sklearn method for logistic regression.
make_blobs	sklearn method to create blobs.
model_selection	sklearn library of methods.
PolynomialFeatures	A method for multivariate regression.
predict	Method of a model for prediction.
r2_score	Method in the sklearn library for squared error.
scikit-learn	A library for data analysis.
sklearn	It calls methods in scikit-learn.
sklearn.metrics	Library of methods to compute metrics.
sigmoid	Activation function for logistic regression.
train_test_split	Method to split data.

11.10   Python Instructions in Chapter 11

The instructions used in Chapter 11 are shown in Table 11.6.

11.11   Conclusions

This chapter has presented a brief introduction to machine learning. Some examples of supervised and unsupervised learning have covered these topics, although not exhaustively, some aspects of this chapter topic. The examples have been made from scratch and also using the package Scikit-learn. There are many more topics in machine learning that are not covered here. The purpose of the chapter is to show the use of Python in machine learning.

Chapter 12

Neural Networks

12.1 Introduction

Artificial neural networks have been around for almost 80 years. They were first proposed in 1943 by McCulloch and Pitts [1]. Neural nets were widely adopted but in the early 1990's their use diminished when it was believed that other new techniques were more promising. It was the development of the increase in computer power, the gaming industry, and the very large amount of data necessary for training neural networks that gave way to further development and use of *artificial neural networks* (ANN). Neural nets were conceived to mimic a biological neuron, as shown in Figure 12.1a. In the figure, it can be seen that a biological neuron cell is composed of a nucleus, input branches known as dendrites, and an output called the axon with axon terminals. Also, in Figure 12.1b it is shown an artificial neuron consisting of the same parts. Nowadays, many researchers have dropped the *artificial* word and the name most commonly used is *neural networks*. The advantage of neural networks is that thousands of them can be used in a network to fulfill a task. In this chapter models for neurons are presented together with the mathematical functions needed. The packages `TensorFlow` and `Keras` are used in the implementation of neural networks. Examples presented show how to use these packages in the implementation of neural networks.

12.2 A Model for a Neuron

Millions of neurons, as the one depicted in Figure 12.1a are interconnected in the brain. Each neuron acts by sending electricity spikes to fire other neurons and to excite other parts of the body. For example, neurons send electricity spikes to a muscle and the effect is to move it to some position. Another example is to convert the sound received by the ear to signals so the brain

DOI: 10.1201/9781003222118-12

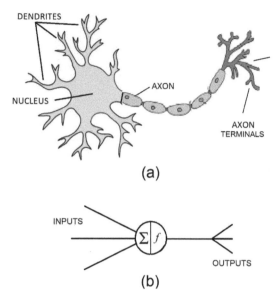

FIGURE 12.1: Biological and artificial neuron.

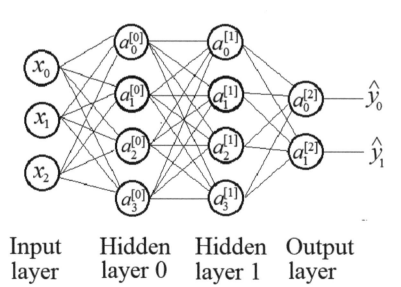

FIGURE 12.2: Typical artificial neural network with inputs and outputs.

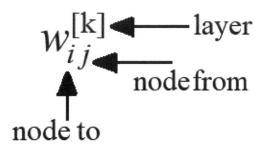

FIGURE 12.3: Indexes for the weights.

interprets the auditory message. These are only two examples of the many activities processed inside the brain by the billions of neurons that compose it. A typical artificial neural network is shown in Figure 12.2. It is formed by inputs, represented by the x_i's, neuron layers consisting of neurons interconnected with the inputs or the outputs of a previous layer, and outputs. The neural network shown in Figure 12.2 has a vector of inputs with three components x_0, x_1, and x_2, two layers of four neurons each, and a vector of outputs with two components. Each neuron is connected to neurons in a previous layer, or to the input vector. The outputs of a previous layer are multiplied by a weight w. There is also an input b called a bias, not shown in this figure. The output of a neuron is denoted by $a_j^{[m]}$ where m is the layer number and j is the neuron position in the layer. The bias coefficient $b_j^{[m]}$ corresponds to the bias in the j neuron in the m layer. The weights are denoted as $w_{ij}^{[m]}$ where j is the source node and i is the end node.

This notation for the weights is shown in Figure 12.3. The model for a neuron is shown in Figure 12.4 and its behavior is to add, represented by the Greek letter Σ, the inputs, and then apply a nonlinear *activation function* represented by the letter f. The addition of the inputs to the jth neuron of the kth layer which are the outputs of the $(k-1)^{\text{th}}$ layer nodes and the bias inputs, which are then

$$z_j^{[k]} = w_{j0}^{[k]} a_0^{[k-1]} + w_{j1}^{[k]} a_1^{[k-1]} + \cdots + w_{j(n-1)}^{[k]} a_{n-1}^{[k-1]} + b_j^{[k]}$$

and then apply an activation function $f(z)$ to obtain

$$f(z_j^{[k]}) = f(w_{j0}^{[k]} a_0^{[k-1]} + w_{j1}^{[k]} a_1^{[k-1]} + \cdots + w_{j(n-1)}^{[k]} a_{n-1}^{[k-1]} + b_j^{[k]})$$

and

$$a_j^{[k]} = f(z_j^{[k]})$$

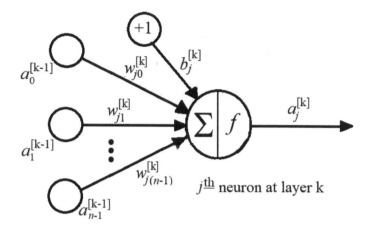

FIGURE 12.4: Model for a neuron with inputs and output.

where $a_j^{[k]}$ is the output of the activation function. The activation function may be the sigmoid function from Chapter 11, but there exist other activation functions as shown later in the chapter. The input $+1$ is called a bias unit and the b's together with the w's are called the weights, also called parameters, which can be represented by column vectors as

$$W^{[k]} = \begin{bmatrix} w_{j0}^{[k]} \\ w_{j1}^{[k]} \\ \vdots \\ w_{j(n-1)}^{[k]} \end{bmatrix} \quad b^{[k]} = \begin{bmatrix} b_{j0}^{[k]} \\ b_{j1}^{[k]} \\ \vdots \\ b_{j(n-1)}^{[k]} \end{bmatrix} \quad z^{[k]} = \begin{bmatrix} z_{j0}^{[k]} \\ z_{j1}^{[k]} \\ \vdots \\ z_{j(n-1)}^{[k]} \end{bmatrix} \quad (12.1)$$

The outputs $a_j^{[k]}$ are also arranged in a vector as

$$a^{[k]} = \begin{bmatrix} a_{j0}^{[k]} \\ a_{j1}^{[k]} \\ \vdots \\ a_{j(n-1)}^{[k]} \end{bmatrix} \quad (12.2)$$

The sums $z_j^{[k]}$ can be written in vector form as

$$z_j^{[k]} = W^{[k]^T} a^{[k]} + b^{[k]}$$

where the T indicates the transpose of the vector. The matrix W is formed by the vectors $W^{[k]}$ as:

$$
W = \begin{bmatrix} W^{[0]^T} \\ W^{[1]^T} \\ \vdots \\ W^{[m-1]^T} \end{bmatrix}
\tag{12.3}
$$

and then the Z and a arrays are formed with

$$
Z = \begin{bmatrix} z^{[0]} \\ z^{[1]} \\ \vdots \\ z^{[m-1]} \end{bmatrix} \qquad
a = \begin{bmatrix} a^{[0]} \\ a^{[1]} \\ \vdots \\ a^{[m-1]} \end{bmatrix}
\tag{12.4}
$$

where a is given by

$$
a = f(Z)
$$

A neural network is a set of neurons interconnected, for example, as shown in Figure 12.2 with multiple inputs, but it may have a single input. The connection between neurons, inputs, and outputs is called the architecture of the neural network. Figure 12.2 shows an example of an architecture. The inputs in this figure are called the input layer, the outputs are called the output layer, and the two layers of neurons that cannot be seen from neither the input nor the output, are called the hidden layers. The figure also shows the names of the output variables for each neuron, namely $a_i^{[j]}$ where j indicates the layer number and i the position of the neuron in the layer. The corresponding equations for the first layer which has four nodes and receives inputs from the vector of inputs X. The equations for the z's in layer 0 are

$$
z_0^{[0]} = w_{00}^{[0]} x_0 + w_{01}^{[0]} x_1 + w_{02}^{[0]} x_2 + b_0^{[0]}
$$

$$
z_1^{[0]} = w_{10}^{[0]} x_0 + w_{11}^{[0]} x_1 + w_{12}^{[0]} x_2 + b_1^{[0]}
$$

$$
z_2^{[0]} = w_{20}^{[0]} x_0 + w_{21}^{[0]} x_1 + w_{22}^{[0]} x_{n-1} + b_2^{[0]}
$$

$$
z_3^{[0]} = w_{30}^{[0]} x_0 + w_{31}^{[0]} x_1 + w_{32}^{[0]} x_2 + b_3^{[0]}
$$

The outputs of the neurons in layer 0 are given by

$$
a_0^{[0]} = f(z_0^{[0]})
$$

$$a_1^{[0]} = f(z_1^{[0]})$$

$$a_2^{[0]} = f(z_2^{[0]})$$

$$a_3^{[0]} = f(z_3^{[0]})$$

These outputs are fed to the neurons in layer 1 and the equations are

$$z_0^{[1]} = w_{00}^{[1]}a_0^{[0]} + w_{01}^{[1]}a_1^{[0]} + w_{02}^{[1]}a[0]_2 + w_{13}^{[1]}a_3^{[0]} + b_0^{[1]}$$

$$z_1^{[1]} = w_{10}^{[1]}a_0^{[0]} + w_{11}^{[1]}a_1^{[0]} + w_{12}^{[1]}a_2^{[0]} + w_{13}^{[1]}a_3^{[0]} + b_1^{[1]}$$

$$z_2^{[1]} = w_{20}^{[1]}a_0^{[0]} + w_{21}^{[1]}a_1^{[0]} + w_{22}^{[1]}a_2^{[0]} + w_{23}^{[1]}a_3^{[0]} + b_2^{[1]}$$

$$z_3^{[1]} = w_{30}^{[1]}a_0^{[0]} + w_{31}^{[1]}a_1^{[0]} + w_{32}^{[1]}a_2^{[0]} + w_{33}^{[1]}a_3^{[0]} + b_3^{[1]}$$

The outputs of the neurons are given by

$$a_0^{[1]} = f(z_0^{[1]})$$

$$a_1^{[1]} = f(z_1^{[1]})$$

$$a_2^{[1]} = f(z_2^{[1]})$$

$$a_3^{[1]} = f(z_3^{[1]})$$

Finally, for the output layer the equations are

$$a_0^{[2]} = f(z_0^{[2]})$$

$$a_1^{[2]} = f(z_1^{[2]})$$

Also, the outputs are

$$\hat{y}_0 = a_0^{[2]}$$

$$\hat{y}_1 = a_1^{[2]}$$

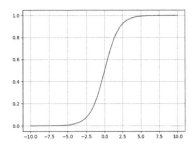

FIGURE 12.5: Sigmoid activation function.

In matrix form

$$a^{[0]} = f(W^{[0]^T} X + b^{[0]})$$

$$a^{[1]} = f(W^{[1]^T} a^{[0]} + b^{[1]})$$

$$a^{[2]} = f(W^{[2]^T} a^{[1]} + b^{[1]})$$

$$\hat{y} = a^{[2]}$$

12.3   Activation Functions

The selection of a proper activation function is of paramount importance. There are several functions used in machine learning. In Chapter 11 the sigmoid function was introduced. Other functions are the **tanh**, the Rectified Linear Unit **RelU**, and the **Leaky RelU**.

The sigmoid function is given by

$$f(z) = \frac{1}{1 + e^{-\alpha z}} \tag{12.5}$$

As it was seen before, it has a value of 0 for $z \ll 0$ and a unity value for $z \gg 0$, as shown in Figure 12.5. It has values between 0 and 1 and it is mostly used in the output layer where the output is a binary one.

The activation function $\tanh(z)$ is plotted in Figure 12.6. It can be seen that the values for this function lie between -1 and +1. This activation function is mostly used in hidden layers of neurons.

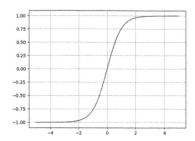

FIGURE 12.6: Hyperbolic tangent activation function.

The Rectified Linear Unit or ReLU, has a linear value for $z > 0$ and it has a 0 value for $z < 0$. It is shown in Figure 12.7 and it is defined by

$$f(z) = \mathtt{max}(0, z) \qquad (12.6)$$

Finally, a variation for the ReLU activation function is the Leaky ReLU that has a small slope for $z < 0$. It is defined by

$$f(z) = \mathtt{max}(0.01z, z) \qquad (12.7)$$

The Leaky ReLU is shown in Figure 12.8. Note the small slope for $z < 0$ which gives the name to this activation function.

It can be shown that the derivatives, used in the cost function, for these activation functions are as follows. For the sigmoid function

$$\frac{d}{dz}f(z) = f(z)[1 - f(z)] \qquad (12.8)$$

For $z \ll 0$ and $z \gg 0$ the derivative $f'(z) \approx 0$. For $z = 0$ the value of $f(0) = 1/4$.

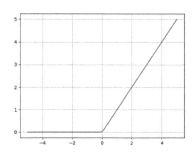

FIGURE 12.7: Rectified Linear Unit $\mathtt{RelU}$ activation function.

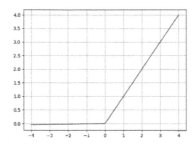

FIGURE 12.8: Leaky `ReLU` activation function.

For the hyperbolic tangent function `tanh`(z) the derivative is

$$\frac{d}{dz}f(z) = 1 - \tanh^2(z) \qquad (12.9)$$

Thus, for $z \ll 0$ and $z \gg 0$ and $f(z) \approx 0$. For $z = 0$ the value of $f(0) = 1$. For the ReLU function, the derivative for $z > 0$ is 1 for $z < 0$ is 0. For the Leaky ReLU, the derivative for $z > 0$ is 1 for $z < 0$ is 0.01.

Finally, if the activation function is the identity function; that is, $f(z) = z$, the neural network implements linear regression.

12.4 Cost Function

By now it is easy to see that a neural network implements logistic regression. The loss function is then defined as

$$\mathcal{L}(\hat{y}, y) = -[y \log\hat{y} + (1 - y) \log(1 - \hat{y})] \qquad (12.10)$$

and the cost function is defined by

$$J(W, b) = \frac{1}{m} \sum_{i=0}^{m-1} \mathcal{L}(\hat{y}^{(i)}, y^{(i)}) \qquad (12.11)$$

which becomes

$$J(W, b) = \frac{1}{m} \sum_{i=0}^{m-1} [-[y^{(i)} \log\hat{y}^{(i)} + (1 - y^{(i)}) \log(1 - \hat{y}^{(i)})] \qquad (12.12)$$

This cost function can be minimized using the gradient descent algorithm of the previous chapter.

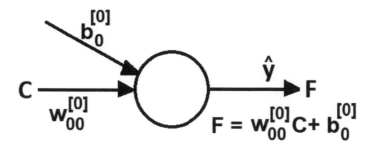

FIGURE 12.9: Single neuron for temperature conversion.

12.5 Tensor Flow

The easiest and fastest way to implement a neural network is to use a library of methods or functions. Several libraries are already available for designing neural networks. The most used ones are `TensorFlow`, developed by Google, and `PyTorch` that was developed by Facebook. This book uses `TensorFlow` to design and execute neural networks. `TensorFlow` includes advanced optimization techniques to search for the parameters that minimize a cost function. It automatically computes the gradients of the functions. `TensorFlow` can be installed using `pip` (see Appendix A). `Keras` is a library for neural nets written in Python that runs over `TensorFlow`. An example that implements linear regression for temperature conversion shows how to use `TensorFlow` and `Keras`.

Example 12.1 Temperature conversion

Consider the case of predicting temperatures at the Fahrenheit scale from Celsius temperatures. The exact equation for the conversion is

$$F = 1.8C + 32 \tag{12.13}$$

Thus for example, for a Celsius temperature of 23°C the corresponding value at the Fahrenheit scale is 73.4°F. As a first neural net consider a single neuron with a single input and a single output, as shown in Figure 12.9. It can be clearly seen that the correct values for b_0 and w_0 are 32 and 1.8, respectively.

To train the neural net, a set of correct equivalent temperatures is given in two arrays as

```
Celsius = np.array([-10, 0, 10, 20, 50, 100])

Fahrenheit = np.array([14.0, 32.0, 50.0, 68.0, 122.0, 212.0])
```

In this example, `Keras` from `TensorFlow` and `numpy` must be imported first:

```
import numpy as np
import tensorflow.keras as ks
```

Keras allows for the design of layers. For the first neural net, only one layer is used. In `Keras` this can be done using the `Dense` layer:

```
layerIn = ks.layers.Dense(units = 1, input_shape = [1])
Out = ks.Dense(units = 1)
model = ks.Sequential([layerIn])
```

Now in the model, the optimizer and the loss criterion have to be defined. The `Adam` optimizer is used with a step of 0.1. A smaller step gives more accuracy but it takes more time and more runs in the learning process. A larger step is not as accurate but is faster for the learning process. The loss criterion is the mean squared error. The model is then:

```
model.compile(optimizer = ks.optimizers.Adam(0.1), \
              loss = 'mean_squared_error')
```

The neural net is ready for training. The parameter `epochs` refers to how many times the training process is done. The training is done with

```
trainingHistory = model.fit(Celsius, Fahrenheit, \
epochs = 1000, verbose = False)
print('Model has been trained')
```

When the training is finished, the message 'Model has been trained' can be printed out. The training history can be plotted with

```
import matplotlib.pyplot as plt
plt.xlabel('No. of epochs')
plt.ylabel('loss')
plt.plot(trainingHistory['loss'])
plt.show()
```

A plot for the training history is shown in Figure 12.10. It can be seen that there are no changes close to the 600th epoch. Now, the neural net is tested with a new Celsius temperature and the result is displayed:

```
print('The neural net is tested')
Ctemp = float(input('Enter a Celsius temperature value: '))
result = model.predict([Ctemp])
print('The result is ' + 'str(result)' + ' F')
```

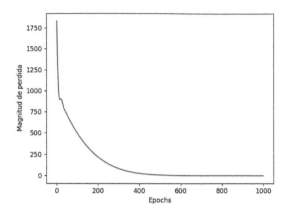

FIGURE 12.10: Learning rate.

The result for a Celsius temperature of 23 °C is 73.30528 which is very close to the correct result of 73.4 °F. The weights in the trained net can be seen with:

```
print('Weights in the model')
print(layerIn.get_weights())
```

The weights are given as:

```
[array([[1.798294]], dtype = float32), array([31.91248],
dtype = float32)]
```

which correspond to $w_0 = 1.798294$ and $b_0 = 31.91248$ and which are very close to the expected values 1.8 and 32. The complete script for this single neuron is:

```
# This is file Example_12.1a

# Keras from TensorFlow and numpy are imported.
import numpy as np
import tensorflow.keras as ks

Celsius = np.array([-10, 0, 10, 20, 50, 100])
Fahrenheit = np.array([14.0, 32.0, 50.0, 68.0, 122.0, 212.0])

# Script continues next page.
```

```
# Script continued from previous page.

# Keras Dense layer is used and the outputs are:
layerIn = ks.layers.Dense(units = 1, input_shape = [1])
Out = ks.layers.Dense(units = 1)

# The model is defined as sequential with:
model = ks.Sequential([layerIn])

# The Adam optimizer is used with a step of 0.1.

model.compile(optimizer = ks.optimizers.Adam(0.1),
loss = 'mean_squared_error')

# The training is done with 1000 epochs.
# When finished display : Model has been trained.
trainingHistory = model.fit(Celsius, Fahrenheit,
 epochs = 1000, verbose = False)
print('Model has been trained')

# Plot the training history.
import matplotlib.pyplot as plt
plt.xlabel('No. of epochs')
plt.ylabel('loss')
plt.plot(trainingHistory['loss'])
plt.show()

# The neural net is tested and the result is displayed.
print('The neural net is tested')
Ctemp = float(input('Enter a Celsius temperature value: '))
result = model.predict([Ctemp])
print('The result is ' + 'str(result)' + ' F')

# The weights in the neural node are:
print('Weights in the model')
print(layerIn.get_weights())
```

The result is: 73.30528 which is very close to the correct result of 73.4 °F. It can be seen that the neural network works correctly.

Next, the neural net is implemented with two more layers with three neurons each which correspond to the hidden layers. The network is shown in Figure 12.11. The layer named `layerIn` is replaced with the following layers:

```
hidden1 = ks.layers.Dense(units = 3, input_shape = [1])
hidden2 = ks.layers.Dense(units = 3)
```

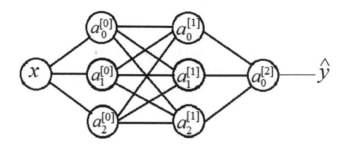

$$\text{Input} \quad \text{Hidden} \quad \text{Hidden Output}$$
$$\text{layer} \quad \text{layer 0} \quad \text{layer 1 layer}$$

FIGURE 12.11: Multilayer neural network.

The output layer is the same. The model is:

```
model = ks.Sequential([hidden1, hidden2, out])
```

The complete script is:

```
# This is file Example_12.1b

# It uses two hidden layers for temperature conversion.

# Import the necessary libraries:
# Keras from TensorFlow and numpy are imported.
import numpy as np
import tensorflow.keras as ks

Celsius = np.array([-10, 0, 10 20, 50, 100])
Fahrenheit = np.array([[14. 32. 50. 68. 122. 212.]])

# Keras Dense layer is used and the output are:
hidden1 = ks.layers.Dense(units = 3, input_shape = [1])
hidden2 = ks.layers.Dense(units = 3)
Out = ks.layers.Dense(units = 1)

# The model is defined as sequential with:
model = ks.Sequential([hidden1, hidden2, Out])

# Script continues next page.
```

```
# Script continued from previous page.

# The Adam optimizer is used with a step of 0.1.
model.compile(optimizer = ks.optimizers.Adam(0.1), \
            loss = 'mean_squared_error')

# The training is done with 1000 epochs.
# When finished, display : Model has been trained.
trainingHistory = model.fit(Celsius, Fahrenheit, \
            epochs = 1000, verbose = False)
print('Model has been trained')

# Plot the training history.
import matplotlib.pyplot as plt
plt.xlabel('No. of epochs')
plt.ylabel('loss')
plt.plot(trainingHistory['loss'])
plt.show()

# The neural net is tested and the result is displayed.
print('The neural net is tested')
Ctemp = float(input('Enter a Celsius temperature value: '))
result = model.predict([Ctemp])
print('The result is ' + str(result) + ' F')

# The weights in the trained net can be seen with:
print('Weights in the model')
print('Weights in the first neural layer:')
print(hidden1.get_weights())
print('Weights in the second neural layer:')
print(hidden2.get_weights())
print('Weights in the output layer:')
print(Out.get_weights())
```

The result for a Celsius temperature of 23°C is 73.32 which is very close to the correct result. The learning process plot is shown in Figure 12.12 and it can be seen that the learning is faster due to the increased number of layers in the neural network. The weights are given as:

```
Weights in the first neural layer:
[array([[-0.77364516, -0.20215535, 0.1994351]],
dtype = float32),
array([-4.726781 , -4.0032654, -2.761373 ],
dtype = float32)]
```

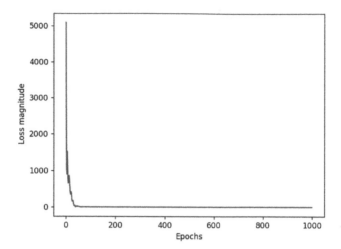

FIGURE 12.12: Learning rate.

```
Weights in the second neural layer:
[array([[ 1.4497128 , 0.18155916, -1.3180187 ],
 [ 0.07405471, 0.32363743, -1.5149053 ],
 [ 0.20846014, 0.89597803, -0.9005167 ]],
dtype = float32), array([-0.9355183 , 0.72436213, 4.323896],
 dtype = float32)]
```

```
Weights in the output layer:
[array([[-0.00702135], [0.38678387], [1.572698]],
dtype = float32), array([3.4024596], dtype = float3
```

Example 12.2 A neural net implementation for the XOR logic gate
The XOR logic gate is implemented in this example. It is implemented with
two hidden layers and is an adaptation of the previous example. In this case,
the activation function is added in the hidden and output layers. The activation function **tanh** is used in the hidden layers and the **sigmoid** function is
used in the output layer. With these changes in mind, the final script is:

```
# This is file Example_12_2.py
# This file computes the XOR logic function.

# Keras from TensorFlow and numpy are imported.
import numpy as np

# Script continues next page.
```

```
# Script continued from previous page.

import tensorflow.keras as ks

# The data is:
X = np.array([[0,0], [0,1.0], [1.0,0], [1.0,1.0]], \
   dtype = np.float32)
Y = np.array([[0], [1.0], [1.0], [0]], dtype=np.float32)

# Keras Dense layer is used and the hidden and output layers
are:
hidden1 = ks.layers.Dense(units = 3, input_shape = [2],
activation = 'tanh')
hidden2 = ks.layers.Dense(units = 3, activation = 'tanh')
Out = ks.layers.Dense(units = 1, activation = 'sigmoid')

# The model is defined as sequential with:
model = ks.Sequential([hidden1, hidden2, Out])

# The Adam optimizer is used with a step of 0.1.
model.compile(optimizer = ks.optimizers.Adam(0.1), \
          loss = 'mean_squared_error')

# The training is done with 1000 epochs.
# When finished, display : Model has been trained.
trainingHistory = model.fit(Celsius, Fahrenheit, \
          epochs = 1000, verbose = False)
print('Model has been trained')

# Plot the training history.
import matplotlib.pyplot as plt
plt.xlabel('No. of epochs')
plt.ylabel('loss')
plt.plot(trainingHistory.history['loss'])
plt.show()

# The neural net is tested and the result is displayed.
result = model.predict(X)
print('The result is: \n', Y)
print('The neural net is tested with an external input')
Xext0 = int(input('Enter an input value: '))%2
Xext1 = int(input('Enter another input value: '))%2
result = model.predict(np.array([[Xext0, Xext1]]))
print('The result is ' + bool(np.rounf(result)))
```

A run with the input [1, 0] produces

```
Model has been trained
1/1 [==============================] - ETA: 0
1/1 [==============================] - 0s 169ms/step

The result is Y:
[[0.02127211]
 [0.9962218 ]
 [0.96937585]
 [0.02036152]]

Enter an input value: 1
Enter another input value: 0

The result is True
```

The results for Y give the probabilities of the result; thus, the output should be read as [0, 1, 1, 0], or [False, True, True, False], which is the correct result.

12.6   Convolutional Neural Networks

Convolutional neural networks are especially efficient to process images. They work by applying a filter, also known as either a mask or a kernel, to an input image. The resulting operation is known as a convolution; thus, the name of the neural network. Recall from Chapter 10 that an image is composed of pixels and that each pixel is represented by an integer number ranging from 0 for black pixels to 255 for white pixels. A mask was applied to an image for morphological transformations.

Now consider the 5×5 image with the pixel values shown in Figure 12.13a and the filter mask, which is a 3×3 mask, shown in Figure 12.13b.

The first step in the procedure to implement the convolution of these two arrays is shown in Figure 12.14a where the mask is superimposed in the image, the pixel values are multiplied and added to produce the (0, 0) pixel in 12.14b.

The next step is to shift the mask one pixel to the right to obtain the next pixel in the convolution as shown in Figure 12.15. The procedure is repeated shifting by another pixel to the right. The result is the pixel (0, 2) with the value 68.

The following step is to move to the left and one pixel downward as shown in Figure 12.16 where the pixel (1, 0) has the value 72.

27	28	34	8	21
43	14	5	12	18
0	0	0	25	6
9	58	17	55	4
7	8	4	20	1

0	1	0
1	1	1
0	1	0

(a)   (b)

FIGURE 12.13: Image with pixel values and filter.

Continuing the procedure of shifting the mask, the final pixel is obtained as shown in Figure 12.17 with the value 121. From this figure it can be seen that the convolved image has been shrinked. It can be shown that the size of the convolved image is given by

$$\text{Size of convolved image} = N - f + 1 \qquad (12.14)$$

where

- N is the dimension of the image.

- f is the dimension of the filter mask.

In digital signal processing, the convolution requires flipping the filter on both the horizontal and vertical axis. For the shown examples, the filter is

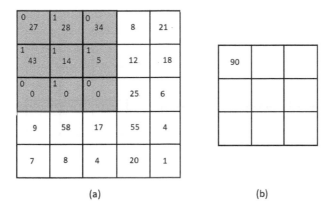

(a)   (b)

FIGURE 12.14: First step in the convolution of the image and the mask.

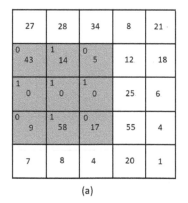

(a)   (b)

FIGURE 12.15: Second step in the convolution of the image and the mask.

symmetrical on both axes, but in other filters, this may be important. As can be seen, what has been implemented is not a convolution but rather a cross-correlation. However, the name convolution is used in neural networks.

12.6.1 Padding

As seen before, a convolution shrinks the image resulting in a smaller size image. Repeating the convolution process several times may produce an image so small that it is not useful for the user. This problem can be solved by padding that consists of adding zeros to the outer rows and columns of the original image. For example, the original image in Figure 12.13a is padded with zeros, as can be seen in Figure 12.18. If p is the number of columns padded to the image (the same number of rows are padded), it can be shown

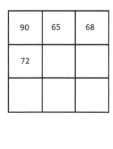

(a)   (b)

FIGURE 12.16: Fourth step in the convolution of the image and the mask.

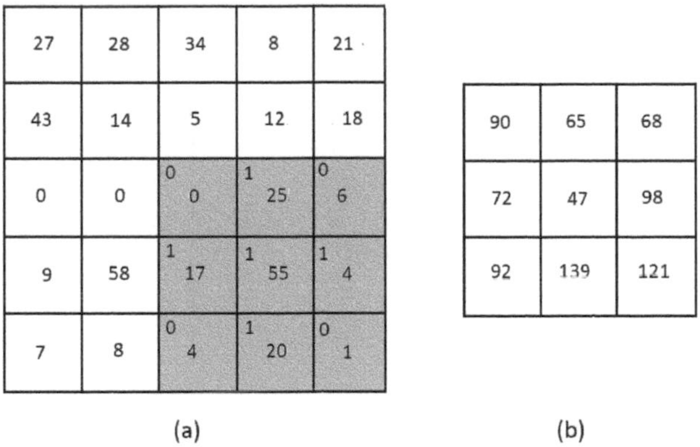

27	28	34	8	21
43	14	5	12	18
0	0	0 / 0	1 / 25	0 / 6
9	58	1 / 17	1 / 55	1 / 4
7	8	0 / 4	1 / 20	0 / 1

90	65	68
72	47	98
92	139	121

(a)   (b)

FIGURE 12.17: Convolution finished.

that the final size of the convolved image is given by

$$\text{Size of convolved image with padding} = N + 2p - f + 1 \qquad (12.15)$$

The image convolved with a 3×3 filter and padded with two columns is

$$\text{Size of convolved image with padding} = 6 + 2 - 3 + 1 = 6$$

which is the same size as the original image. In addition, to preserve the size of the image, padding also permits that the pixels at the edges contribute more in the convolution because they are taken into account more often than before. In addition, there is a central pixel in every step of the convolution

0	0	0	0	0	0	0
0	27	28	34	8	21	0
0	43	14	5	12	18	0
0	0	0	0	25	6	0
0	9	58	17	55	4	0
0	7	8	4	20	1	0
0	0	0	0	0	0	0

FIGURE 12.18: Image with zero-padding.

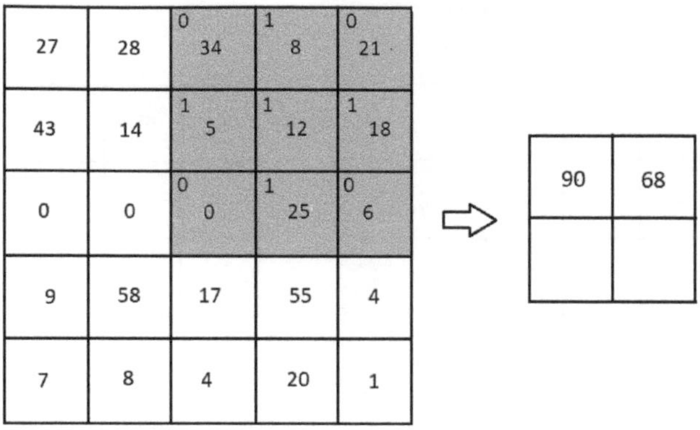

FIGURE 12.19: Second step of a strided convolution.

and this is the reason that every pixel in the image, either on the edges or in the inner parts of the image, contribute to the convolution result.

12.6.2   Strided convolution

In a strided convolution the filter is shifted more than one column at a time. Consider Figure 12.13. The first step in the convolution is computed as before and shown in Figure 12.14. The second step is computed now by shifting two columns at a time as shown in Figure 12.19. This step produces the pixel (0, 1) with the value 68. Notice that the right end of the image has been reached and thus the next step in the process is to start at the left end but two rows below. The last step is to shift the filter two columns to the right and this produces Figure 12.20 with the resulting convolution with a size 2 × 2. It can be shown that the resulting image size is given by

$$\text{Size of the convolved image with striding} = \frac{N - f}{s} + 1 = 6$$

where s represents the stride size. Thus, for the image shown in Figure 12.13 which is a 5×5 image, N is 5, the filter size is $f = 3$, and $s = 2$. Thus,

$$\text{Size of the convolved image with striding} = \frac{5 - 3}{2} + 1 = 2 \qquad (12.16)$$

12.6.3   Pooling

A form to reduce the size of the processed image is the use of pooling. This is a technique that divides the image in equal-size parts and then implement any of two techniques, namely max-pooling and mean-pooling.

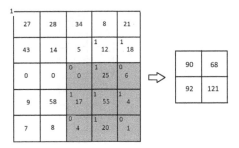

FIGURE 12.20: Last step of a strided convolution.

Max-pooling is shown in Figure 12.21. There the image is partitioned in 2×2 sections (Figure 12.21a) and the maximum value in each section is taken into the max-pooled image as it can be seen in Figure 12.21b. In this Figure a stride of 2 and a filter size 2×2 are used.

Mean-pooling, also known as average-pooling, is implemented by taking the average or mean value of the element values covered by the mask. Figure 12.21c shows the result from mean-pooling.

12.7   A Layer of a Convolutional Filter

A color image has three components corresponding to the three colors red, green and blue. Each component can be filtered by three different filters

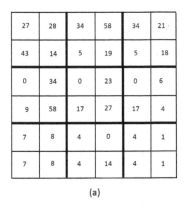

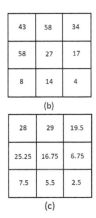

FIGURE 12.21: Pooling of an image. a) Original image, b) result for max-pooling, c) Result for mean-pooling.

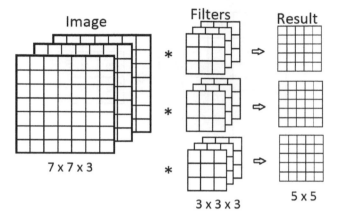

FIGURE 12.22: Convolution for a color image.

as shown in Figure 12.6. The convolution of each filter produces a two-dimensional image composed by the sum of the corresponding convolutions for each filter. This process is shown in Figure 12.22 where the asterisk denotes the convolution. For each filter, there is a result, which in this example is a 5×5 image. Each resulting image has some image characteristic that was obtained using an adequate filter.

The next step is to apply a non-linearity as shown in Figure 12.23. This step can use any of the known activation functions, as can be seen in this figure. The arrangement shown is called a layer of a convolutional neural network.

The variables involved in the process described in Figure 12.23 are as follows:

- The input X for this example has 7×7×3 pixels. In the case of a larger image size; for example, 1024×1024 the image size is 3,145,728 pixels. For a layer which is not in the input, the input is the output of a

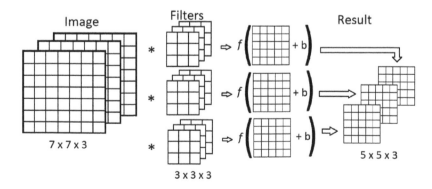

FIGURE 12.23: Layer of a convolutional neural network.

previous layer which is $a^{[j]}$. So for the first convolutional layer, the input corresponds to $a[-1]$.

- The filters are of size $3 \times 3 \times 3$ pixels that gives 27 pixels for each filter. If the total number of filters is 10 then the total for the filters is 270. The output of the filters are the parameters $z[j]$.

- The output of the filters is an array of size 5×5 for this example. If the size of the original image is 1024×1024 then the output size is 1022×1022. To this array add the bias coefficient b and then apply the activation function f. The addition of the bias coefficient adds 1 to the weight parameters. This gives a total of 28 parameters per filter.

- The activation function is applied to the output of the filters plus the bias coefficient. The output of the activation function is $a^{[j+1]} = f(z[j])$. It is a 5×5 array for each filter.

- The total number of parameters for each filter is 27 weights and the bias for a total of 28 parameters for each filter.

- The last step in the neural net consists of the flattening of the arrays to form a one-dimensional vector which is applied to a regular neural layer.

- Some times, a pooling process is applied after a convolution layer.

As an example, Figure 12.24 shows a convolutional neural network with two convolutional layers. The image to be processed is a three-dimensional image with size of $21 \times 21 \times 3$. The first convolutional layer consists of three filters, neither striding nor pooling, and ten filters. The result is a set of 10 arrays with a size of $19 \times 19 \times 10$. The next step is the application of a second convolutional layer with five filters, a stride of 2, no pooling, and 20 filters. The result is a set of 20 arrays with size of 7×7. The next step is a flattening which consists of forming a one-dimensional vector from the elements in the arrays. A neural net layer receives this vector and the result is sent to a softmax activation function to make a decision.

The convolutional neural network is implemented with the Keras method Conv2D. The arguments used here for this method are:

- The number of filters.

- The kernel size.

- The strides.

- The padding.

- The activation.

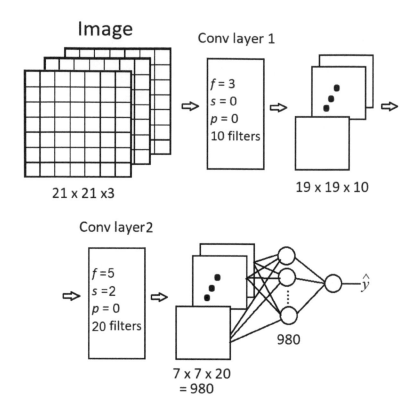

FIGURE 12.24: Example of a convolutional neural network.

Example 12.3 A convolutional neural network to classify images
Consider the MNIST fashion data. This is a set of 70,000 clothing and shoes images divided in 10 different image types. The set is available in the Tensor-Flow datasets. It includes the images, the metadata, and other data, such as the number of items, the labels, etc. The neural network to classify the data is a convolutional neural network implemented with the Conv2D method from Keras. Besides the `Conv2D` method, the `MaxPooling2D` method is also used. After the convolutional filters are applied, the image is flattened which means that the rows in the array are aligned to form a vector. Then a 100 neurons layer is applied, and finally, a neuron with the `softmax` activation function is applied. This example is run in **Google Colab**. To start the script, import the necessary libraries:

```
import tensorflow as tf
import tensorflow_datasets as tfds
```

The Fashion MNIST dataset is downloaded:

```
# Download data for Fashion MNIST from Zalando
data, metadata = tfds.load('fashion_mnist', \
        as_supervised = True, with_info = True)
```

The metadata is printed out to see the information it contains, and it is seen that name, description, amount of data, type of the items, homepage, size, keys, images and labels, and the citation.

```
tfds.core.DatasetInfo(
name = 'fashion_mnist',
full_name = 'fashion_mnist/3.0.1',
description = """
Fashion-MNIST is a dataset of Zalando's article images
consisting of a training set of 60,000 examples and a test set
of 10,000 examples. Each example is a 28x28 grayscale image,
associated with a label from 10 classes.""",

homepage = 'https://github.com/zalandoresearch/fashion-mnist',
data_dir = '/root/tensorflow_datasets/fashion_mnist/3.0.1',
file_format = tfrecord,
download_size = 29.45 MiB,
dataset_size = 36.42 MiB,
features = FeaturesDict({
'image': Image(shape=(28, 28, 1), dtype=uint8),
'label': ClassLabel(shape=(), dtype=int64, num_classes=10), }),
supervised_keys = ('image', 'label'),
disable_shuffling = False,
splits = {
'test': <SplitInfo num_examples = 10000, num_shards = 1>,
'train': <SplitInfo num_examples = 60000, num_shards = 1>, },

citation = """@ article DBLP:journals/corr/abs-1708-07747,
author = {Han Xiao and
Kashif Rasul and
Roland Vollgraf},
title = {Fashion-MNIST: a Novel Image Dataset for
Benchmarking Machine
Learning Algorithms},
journal = {CoRR},
volume = abs/1708.07747},
year = {2017},
url = {http://arxiv.org/abs/1708.07747},
archivePrefix = {arXiv},
```

```
eprint = {1708.07747},
timestamp = {Mon, 13 Aug 2018 16:47:27 +0200},
biburl = {https://dblp.org/rec/bib/journals/corr/abs-1708
-07747}, bibsource = {dblp computer science bibliography,
https://dblp.org}""",
)
```

Next, the data is separated into the training data and the testing data:

```
# Data is separated in training data and testing data:
training_data, testing_data = datos['train'], datos['test']
```

The label names can be printed out as:

```
# Label names:
label_names = metadata.features['label'].names
print('The label names are: ', label_names)
```

and the label names are:

['T-shirt/top', 'Trouser', 'Pullover', 'Dress', 'Coat', 'Sandal',
'Shirt', 'Sneaker', 'Bag', 'Ankle boot']

and it can readily be seen that there are 10 item types. As can be seen in the metadata, the images are arrays with a size of 28×28×1; that is, they are gray level images with a type uint8. They are 8-bit numbers with gray-level values between 0 and 255. They have to be normalized to have values between 0 and 1. First, they have to be converted or cast to floating type and then normalized. This is done in the function normalize as:

```
def normalize(images, labels):
    images = tf.cast(images, tf.float32)
    images /= 255 # The normalization is done here.
    return images, labels
```

This function is applied to both the training data and the testing data and then they are put in the cache memory to speed up the process:

```
training_data = training_data.map(normalize)
testing_data = testing_data.map(normalize)
# Add to cache memory to speed up the process:
training_data = training_data.cache()
testing_data = testing_data.cache()
```

The model is created with two convolutional networks Conv2D, two pooling layers MaxPooling2D, a flattening layer Flatten, and a 100-neuron layer. The model is:

```
model = tf.keras.models.Sequential([
tf.keras.layers.Conv2D(32, (3,3), activation ='relu' \
, input_shape = (28, 28, 1)),
tf.keras.layers.MaxPooling2D(2, 2),

tf.keras.layers.Conv2D(64, (3,3), activation ='relu'),
tf.keras.layers.MaxPooling2D(2,2),

tf.keras.layers.Flatten(),
tf.keras.layers.Dense(100, activation ='relu'),
tf.keras.layers.Dense(10, activation = "softmax")
])
```

The model is compiled as:

```
model.compile(optimizer = 'adam',
loss = tf.keras.losses.SparseCategoricalCrossentropy
(from_logits = True), metrics = ['accuracy'])
```

The training and testing data examples are split:

```
training_ex_number = metadata.splits["train"].num_examples
testing_ex_number = metadata.splits["test"].num_examples
```

The fitting of the model is done in batches with a lot size of 64 with:

```
LOT_SIZE = 64
training_data = training_data.repeat(). \
shuffle(training_ex_number).batch(LOT_SIZE)
testing_data = testing_data.batch(LOT_SIZE)
```

Now, the training is done with 2 epochs:

```
# Training
import math
Model = model.fit(training_data, epochs = 2, \
steps_per_epoch = math.ceil(training_ex_number/LOT_SIZE))
```

The output from the fitting is:

```
Epoch 1/2
938/938 [============] - 61s 57ms/step - loss: 0.4995 -
accuracy: 0.8208
Epoch 2/2
938/938 [============] - 52s 56ms/step - loss: 0.3312 -
accuracy: 0.8794
```

After the two epochs are finished, the loss has been reduced to 0.3312 and the accuracy is at 0.8794. These results can be improved by increasing the number of epochs. Now, some of the elements in a batch are plotted together with the predictions computed. This is done with two methods as follows:

```python
def plot_image(i, arr_predictions, real_labels, images):
    arr_predictions, real_label, img = arr_predictions[i], \
    real_labels[i], images[i]
    plt.grid(False)
    plt.xticks([])
    plt.yticks([])
    plt.imshow(img[...,0], cmap = plt.cm.binary)
    prediction_label = np.argmax(arr_predictions)
    if prediction_label == real_label:
        color = 'blue'
    else:
        color = 'red'
    plt.xlabel("{}{:2.0f}% ({})".format(class_names \
    [prediction_label],
        100*np.max(arr_predictions), class_names[real_label]),\
        color = color)

def plot_value_array(i, arr_predictions, real_label):
    arr_predictions, real_label = arr_predictions[i],\
     real_label[i]
    plt.grid(False)
    plt.xticks([ ])
    plt.yticks([ ])
    plot1 = plt.bar(range(10), arr_predictions, \
     color="# 777777")
    plt.ylim([0, 1])
    label_prediction = np.argmax(arr_predictions)
    plot1[label_prediction].set_color('red')
    plot1[real_label].set_color('blue')
```

FIGURE 12.25: Fashion images with plots of accuracy.

```
import matplotlib.pyplot as plt
rows = 10
columns = 5
num_images = rows*columns
plt.figure(figsize = (2*2*columns, 2*rows))
for i in range(num_images):
    plt.subplot(rows, 2*columns, 2*i+1)
    plot_image(i, predictions,testing_labels,testing_images)
    plt.subplot(rows, 2*columns, 2*i+2)
    plot_value_array(i, predictions, testing_labels)

plt.show()
```

The plots together with the predictions are shown in Figure 12.25. Finally, the model is tested with a single image:

```
# Test with a single image
image = testing_images[19] # Test image from the plot.
image = np.array([image])
prediction = Model.predict(image)
print("Prediction: " + class_names[np.argmax(prediction[0])])
```

For the testing image No. 19 the result is:

```
Prediction: Pullover
```

The model summary can be printed out with:

```
Model.summary()
```

The summary is:

Layer (type)	Output Shape	Param #
conv2d_10 (Conv2D)	(None, 26, 26, 32)	320
max_pooling2d_10 (MaxPooling2D)	(None, 13, 13, 32)	0
conv2d_11 (Conv2D)	(None, 11, 11, 64)	18496
max_pooling2d_11 (MaxPooling2D)	(None, 5, 5, 64)	0
flatten_5 (Flatten)	(None, 1600)	0
dense_10 (Dense)	(None, 100)	160100
dense_11 (Dense)	(None, 10)	1010

```
Total params: 179926 (702.84 KB)
Trainable params: 179926 (702.84 KB)
Non-trainable params: 0 (0.00 Byte)
```

This summary displays the trainable parameters. The complete file is:

```python
# This is file Example_12_3.py
# It implements a convolutional neural network.

# Import the necessary libraries from TensorFlow
import tensorflow as tf
import tensorflow_datasets as tfds

# Download the data Fashion MNIST from Zalando.
datos, metadata = tfds.load('fashion_mnist',
as_supervised=True, with_info=True)

# Display the metadata:
print(metadata)

# Script continues next page.
```

```
# Script continued from previous page.

# Split the data into a training set (60K) and
# a testing set (10K).
training_data, testing_data = datos['train'], datos['test']

# Labels of the 10 categories
class_names = metadata.features['label'].names

# Display the class names:
print(class_names)

# Data normalization (Values from 0-255 to 0-1)

def normalize(images, labels):
    images = tf.cast(images, tf.float32)
    images /= 255 # Pixel values are between 0-1.
    return images, labels

# Actual normalization is done:
training_data = training_data.map(normalize)
testing_data = testing_data.map(normalize)
# Add to cache a faster training.
training_data = training_data.cache()
testing_data = testing_data.cache()
# print(training_data.cache())

# Create the model:
Model = tf.keras.models.Sequential([
tf.keras.layers.Conv2D(32, (3,3), activation='relu', \
input_shape = (28, 28, 1)), \
tf.keras.layers.MaxPooling2D(2, 2),

tf.keras.layers.Conv2D(64, (3,3), activation = 'relu'),
tf.keras.layers.MaxPooling2D(2,2),
tf.keras.layers.Flatten(),
tf.keras.layers.Dense(100, activation = 'relu'),
tf.keras.layers.Dense(10, activation = "softmax")])

# Compile the model
Model.compile(optimizer = 'adam',
loss = tf.keras.losses.SparseCategoricalCrossentropy \
(from_logits = True), metrics = ['accuracy'])

# Script continues next page.
```

```
# Script continued from previous page.

# Training and testing data split:
training_ex_number = metadata.splits["train"].num_examples
testing_ex_number = metadata.splits["test"].num_examples

# Make batches for a more efficient learning:
LOT_SIZE = 64

# Shuffle mixes the data so the neural net does not learn
# for a fixed order:
training_data = training_data.repeat(). \
shuffle(training_ex_number).batch(LOT_SIZE)
testing_data = testing_data.batch(LOT_SIZE)

# Training:
import math
history = Model.fit(training_data, epochs=1, \
  steps_per_epoch= math.ceil(training_ex_number/LOT_SIZE))

# Plot a few images:
import numpy as np

def plot_image(i, arr_predictions, real_labels, images):
    arr_predictions, real_label, img = arr_predictions[i], \
    real_labels[i], images[i]
    plt.grid(False)
    plt.xticks([])
    plt.yticks([])
    plt.imshow(img[...,0], cmap = plt.cm.binary)
    prediction_label = np.argmax(arr_predictions)
    if prediction_label == real_label:
        color = 'blue'
    else:
        color = 'red'
    plt.xlabel("{}{:2.0f}% ({})".format(class_names \
    [prediction_label],
        100*np.max(arr_predictions), class_names[real_label]),\
        color = color)

def plot_value_array(i, arr_predictions, real_label):
    arr_predictions, real_label = arr_predictions[i], \
     real_label[i]
    plt.grid(False)

# Script continues next page.
```

```
# Script continued from previous page.

    plt.xticks([])
    plt.yticks([])
    plot1 = plt.bar(range(10), arr_predictions, color =\
     "# 777777")
    plt.ylim([0, 1])
    label_prediction = np.argmax(arr_predictions)
    plot1[label_prediction].set_color('red')
    plot1[real_label].set_color('blue')

import matplotlib.pyplot as plt
rows = 10
columns = 5
num_images = rows*columns
plt.figure(figsize = (2*2*columns, 2*rows))
for i in range(num_images):
    plt.subplot(rows, 2*columns, 2*i+1)
    plot_image(i, predictions, testing_labels, testing_images)
    plt.subplot(rows, 2*columns, 2*i+2)
    plot_value_array(i, predictions, testing_labels)
plt.show()

# Test with a single image
image = testing_images[19] # Only images in the test batch..
image = np.array([image])
prediction = Model.predict(image)
print("Prediction: " + class_names[np.argmax(prediction[0])])

# The model summary is printed out with:
Model.summary()
```

12.8   Python Instructions in Chapter 12

Table 12.1 shows the instructions used in Chapter 12.

TABLE 12.1:   Python instructions in Chapter 12

Instruction	Description
Adam	Optimization algorithm.
compile	It compiles the model.
Conv2D	It implements a 2-dimensional convolution.
fit	It implements the fitting of an algorithm.

(*Continues next page.*)

TABLE 12.1: Python instructions in Chapter 11 (*Continued*)

Instruction	Description
Dense	Fully connected neural network.
epochs	Times the fitting is performed.
Flatten	It converts a matrix to a vector.
features	Parameters of the machine learning problem.
Keras	Library of methods for machine learning.
Leaky Relu	Activation function.
layers	Part of a neural network.
MaxPooling2D	It chooses the maximum of a subimage.
mean-squared-error	It computes the mean-squared error.
MNIST	Library of data
Relu	Activation function.
Sequential	It indicates the ordering of the layers.
TensorFlow	Library of methods for machine learning.
tanh	Activation function.
softmax	Activation function.

12.9 Conclusions

This chapter has presented an introduction to neural networks for machine learning. The examples have been used for classification. The use of an activation function in the neural networks is used to apply non-linearities to the neurons and decide if a given neuron is activated. Different activation functions were used in the examples. Multilayer neural networks were used in the examples and convolutional neural networks were used to classify images. The examples were solved using TensorFlow and Keras which provide the methods to implement many of the needs in neural network design and implementation. Besides being a very complete library of methods, they are both free software, also called open source and can be downloaded and used without restrictions. Thus, the chapter provides an introduction to the topic and the use of Python for the design and implementation of neural networks and it does not pretend to be a comprehensive chapter on the topic.

Appendix A

Installation of Libraries and Packages and Running the Scripts

A.1   Introduction

This appendix provides the necessary steps to run the scripts in the book using Google Colab and to install Python as well as the packages and libraries needed for the book. Instructions are given for the Windows and Mac OS operating systems.

A.2   Running a Script using Google Colab

Google Colab is an interactive environment to run Python scripts in a notebook style. It does not need to be installed and it is ready to use. It can be accessed from the following page:

```
https://colab.research.google.com/?authuser = 1
```

The welcome page is shown in Figure A.1. There are two keywords useful for the user and they are located above the Welcome message. They are +Code and +Text. Pressing the word +Code, a cell is open, as shown in Figure A.2 and there the script can be written. The script (or the cell) can be run by pressing the play symbol to the left of the cell. An advantage of Google Colab is that most of the packages used in the book are already installed. The interested readers can find more information on the welcome window.

A.3   Installation of Python

Python is a free distribution software. To install Python it is necessary to download it first and then install it. Please follow the following steps to install it depending upon the platform.

DOI: 10.1201/9781003222118-A

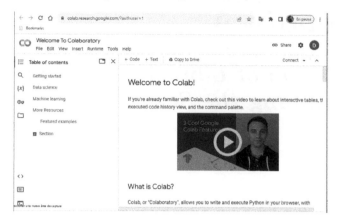

FIGURE A.1: Welcome screen for Google Colab.

Windows

1. The Windows operating system in modern computers is a 64-bit architecture. To check the system architecture go to the `Windows Start` icon, open `Configuration`, choose `System` and `About...`, in the Type of system the architecture is displayed.

2. Open the website https://www.python.org/ and download the latest version.

3. Select the latest Python version, 3.12.0 at the time of this writing, and select the appropriate system, either a 32-bit or a 64-bit architecture.

4. When the download is finished, click on the installation file. Select the option: "`Customize installation`", as shown in Figure A.3.

5. Figure A.4 shows the Optional Features window. Select all the options. Note that the pip feature is going to be installed. Then choose `Next`

6. In the next window select the installation folder as `C:\Python3x\` and continue with the default options. (A different folder can be chosen though). (See Figure A.5).

FIGURE A.2: A cell in Google Colab.

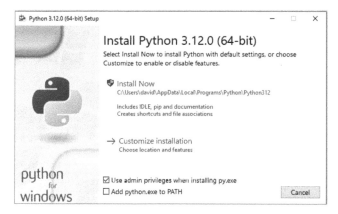

FIGURE A.3: Selecting the option: "`Customize installation`".

7. To check if the installation was successful, run the Python IDLE from the Windows Start menu or from the Search window.

Mac OS X

1. To verify if the operating system is either a 32-bit or a 64-byt architecture, from the Apple icon in the upper left corner choose **About this Mac**, and then choose the processor. If the processor is a core or core duo, the system is a 32-bit system. Otherwise, it is a 64-bit system.

2. Open the website https://www.python.org/ and download the latest version, 3.12.0 at the time of this writing.

3. To check if the installation was successful, go to **Applications** and click on the Python IDLE. If the IDLE window opens then everything is OK.

FIGURE A.4: Selecting the options.

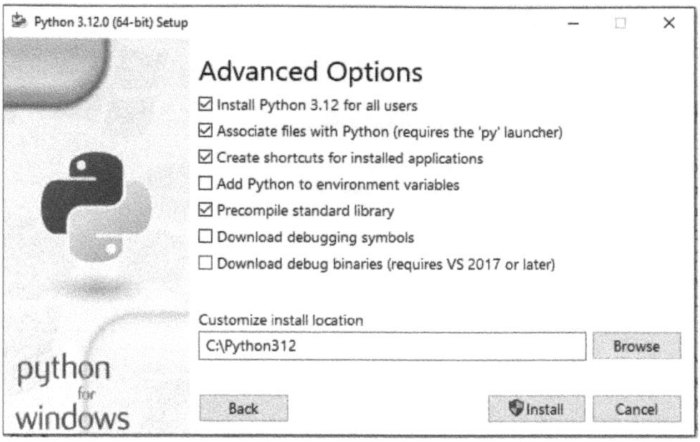

FIGURE A.5: Selecting the installation directory and other options.

A.4 Modules and Libraries Installation with pip

To install modules and libraries provided by third parties, the basic Python installation includes a program known as `pip`. The name `pip` is an acronym for "`pip install packages`".

pip in Windows

1. In the `C:` directory (or in the installation directory) there is a folder Python312. Inside this directory there is a folder named `Scripts`. Inside this folder there are several pip files. The most recent one is `pip3.exe`. If this is the case then `pip` is all set.

2. Open a command window and change directory to C:\python312\scripts and execute:

   `pip3 install <Enter >`

3. The command answer must be:

   /

   ERROR: You must give at least one requirement to install (see "pip help install")

   This is an indication that the pip files are installed.

pip in Mac OS X/ Unix y Linux

1. The pip files are installed when Python is installed.

2. To check that the pip files are ready to install modules or libraries, open a Terminal window and type

   ```
   pip install <Enter >
   ```

   If the system returns:

   ```
   You must give at least one requirement to install
   ```

   Then the pip files are installed and ready to run.

3. It might be the case that there are several Python installations in the same computer. In that case the installation must include the Python version as:

   ```
   python3.12 -m pip install numpy
   ```

A.5   Installation of Modules or Libraries

To install additional modules or libraries it is very easy and the pip files are used for this purpose. An example shows the procedure.

For example, to install the Numpy module with plenty of mathematical functions the required step are shown below for the three platforms.

- **Windows**

  1. Open a command window. Change directory to `C:\python312\scripts` and type:

     ```
     pip3 install numpy <Enter >
     ```

- Open the `IDLE` and execute:

```
>>>  import numpy
```

If there is no error message then **numpy** has been installed properly.

- In the same way install any additional package.

- The command window can be closed now.

Mac OS X/ Unix y Linux

1. Open a Terminal window or shell and execute

```
sudo pip install numpy
```

2. Open the IDLE and execute:

```
import numpy
```

If there is no error message, then **numpy** has been installed properly.

3. In the same way install any additional package.

A.6   Installation of OpenCV

The installation of OpenCV can be done using pip. It can be installed in both Windows and Mac. It is better to run OpenCV scripts in the Python IDLE.

A.6.1   Installation in Windows for the IDLE

Open a Terminal window by typing `cmd` in the search window next to the Windows icon. There type

```
pip install opencv-contrib-python
```

The computer then downloads the required files. When the installation is finished open the Python IDLE and type

```
>>> import cv2
```

If the IDLE shows the Python prompt then OpenCV has been successfully installed.

A.6.2   Installation in the Mac for the IDLE

In the Mac open a Terminal window in Applications -> Utilities -> Terminal and there type

```
sudo pip install opencv-contrib-python
```

The computer then downloads the required files. When the installation is finished open the Python IDLE and type

```
>>> import cv2
```

If the IDLE shows the Python prompt then OpenCV has been succesfully installed.

Index